Praise for

Bokashi Composting

I have to thank you for writing this book Adam — FINALLY we have a well-researched, comprehensive guide to bokashi composting. I've already changed a few of my techniques based on your advice and my bokashi is decomposing faster as a result. Your information is thorough and your writing style is clear and refreshingly humble. You've really created THE step-by-step guide for making bokashi, and people lucky enough to pick up this book will have created a beautiful microbial inoculant for their gardens and houseplants in no time.

> — Phil Nauta, www.SmilingGardener.com,
> author of *Building Soils Naturally*

As we reach our teens in the 21st Century, it's clear that we must explore more and newer ways of reducing our waste output. We have already achieved great reductions in what enters urban landfill sites, and more efficient recycling. But what can we do in our own homes — in our urban condos and apartment buildings? In this upbeat, informative book, Adam Footer reveals the bokashi composting option, and shows us how to further reduce our food waste, turning it back into the Earth and enriching the soil as we do. His explanation of bacterial culturing is clear and simply stated. Whether you purchase a home bokashi kit or build your own, following Footer's foolproof instructions, this is the manual for you.

> — Mark Macdonald, West Coast Seeds

BOKASHI
COMPOSTING

ADAM FOOTER

BOKASHI COMPOSTING

SCRAPS to SOIL in WEEKS

new society
PUBLISHERS

Cover design by Diane McIntosh.
Images © iStock : top veg esp-imaging;
bokashi composted veg – author; soil – craftvision

Printed in Canada. First printing November 2013.

New Society Publishers acknowledges the financial support of
the Government of Canada through the Canada Book Fund (CBF)
for our publishing activities.

Paperback ISBN: 978-0-86571-752-7
eISBN: 978-1-55092-549-4 (ebook)

Inquiries regarding requests to reprint all or part of *Bokashi Composting*
should be addressed to New Society Publishers at the address below.

To order directly from the publishers, please call toll-free (North America)
1-800-567-6772, or order online at www.newsociety.com

Any other inquiries can be directed by mail to:
New Society Publishers
P.O. Box 189, Gabriola Island, BC V0R 1X0, Canada
(250) 247-9737

New Society Publishers' mission is to publish books that contribute in funda-
mental ways to building an ecologically sustainable and just society, and to do so
with the least possible impact on the environment, in a manner that models this
vision. We are committed to doing this not just through education, but through
action. The interior pages of our bound books are printed on Forest Stewardship
Council®-registered acid-free paper that is **100% post-consumer recycled**
(100% old growth forest-free), processed chlorine-free, and printed with vege-
table-based, low-VOC inks, with covers produced using FSC®-registered stock.
New Society also works to reduce its carbon footprint, and purchases carbon off-
sets based on an annual audit to ensure a carbon neutral footprint. For further
information, or to browse our full list of books and purchase securely, visit our
website at: www.newsociety.com

Library and Archives Canada Cataloguing in Publication

Footer, Adam, author

Bokashi composting : scraps to soil in weeks / Adam Footer.

Includes index.

ISBN 978-0-86571-752-7 (pbk.)

1. Compost. 2. Organic fertilizers. I. Title.

S661.F66 2014   631.8'75   C2013-905773-0

Dedication

To my wife Kim, for her undying support in all my endeavors, and to my daughters Kelly and Nicole, for showing me that the future is worth saving.

Contents

Introduction 1

1: Why Bokashi? 7

2: The History of Bokashi 23

3: The Science 29

4: How to Make Bokashi Bran 45

5: The Fermentation Vessel and How to Make Your Own 73

6: How to Compost Your Kitchen Waste with Bokashi 89

7: Using the Fermented Food Waste 101

8: Bokashi Leachate 119

Conclusion 125

Appendix A: Bokashi FAQs 129

Appendix B: Case Studies 141

Appendix C: Further Reading 151

Appendix D: Works Cited and Notes 153

Index.. 157

About the Author.. 163

Introduction

THE GOAL OF THIS BOOK IS TO RAISE AWARENESS OF BOKASHI as a legitimate form of composting. The roots of bokashi composting lie in Japan and Southeast Asia, where it is widely practiced. There is some more recent history and minor usage in countries like Australia, the UK, New Zealand, and South Africa, but a relatively small amount of information has made its way over to the US up to this point. Most Americans don't compost organic waste regularly, and the majority of those who do have never heard of bokashi, so the whole concept of fermenting food waste using bokashi is unheard of for the most part. Yet the benefits of bokashi composting are profound, so it is worth getting the information out there and spreading the word.

Bokashi is a form of composting that uses a specific group of microbes to anaerobically ferment organic matter, resulting in a finished product that can be rapidly digested by the soil biota. The process doesn't require mixing of greens and browns and doesn't generate heat or greenhouses gasses, and all of the by-products are contained within a closed system so nutrients aren't lost in the composting process. Since the system is closed, the user doesn't have to worry about insect or rodent problems, or unpleasant odors emanating from a pile of kitchen waste. All of these advantages make bokashi a good option for someone with space constraints. That might be an apartment dweller, an occupant of an office building, or anyone who doesn't have room for a large traditional compost pile. If you have enough room for a few five-gallon buckets, then you can compost all of your kitchen waste, keeping it out of the landfill and ending up with a finished product that will add a lot of organic matter to your garden. Bokashi composting is also a potential solution for individuals who have tried to compost organic waste in the past using more traditional techniques but have been unsuccessful for one reason or another. The bokashi composting process takes a lot of the complexity out of composting food waste, making the whole process much easier to follow for the average person, so hopefully they can recycle all of their kitchen waste.

When I decided to write this book, I wanted to be sure that I didn't extol the virtues of bokashi composting at the expense of all of other forms of composting. There are absolutely times where vermicomposting, traditional hot composting, or windrow composting are appropriate and/or necessary if we want to keep a lot of organic waste out of the landfill. Each form of composting has its appropriate place and time, but there are also times when one form or another may be problematic given the user and their specific situation. Each form of composting also has its distinct set of disadvantages, so given our own unique situations, it is important that we have a variety of ecological methods to choose from so we can recycle all of our organic wastes. I encourage everyone to compost or reuse all of their organic wastes and keep all (or as much as they can) of their organic material on their property. Bokashi composting may work well for some in achieving that goal so it deservers to be shared and practiced by a wider audience.

A final note. When I first started to do research for this book, it quickly became apparent that information about bokashi was out there, but it was scattered and often very brief. A lot of the same information is repeated over and over. There is a fair amount of good information out there, but not nearly enough, and the good information is hard to find. So I have tried my absolute best to consolidate as much of the quality information about bokashi composting into one place, making it easy for anyone to access and start composting using bokashi. As there are more and more bokashi practitioners, each will have their own experiences and questions will arise. From there more research can be done, and the information and techniques involved with bokashi composting can be expanded and perfected over time. There isn't a lot of quality information or studies out there about bokashi composting, so I hope that practitioners will become inspired to start composting using bokashi and to push the current limits, to find better and more efficient ways to do things. As I previously stated, bokashi composting is just one piece of the puzzle. If we can continue making the recycling of organic wastes easier and more efficient, more and more people who didn't previously compost will start composting. And more composting means that less organic waste is going the landfill, and more stable organic matter is going back into the soil where it belongs.

This book serves as a introductory practical manual about bokashi composting. Anyone reading it will get a basic introduction to the topic and will then be able to start composting their kitchen waste with bokashi. To help simplify things and make the reading easier I need to lay out some basic terminology that I will be using throughout the book. A number of the terms have analogous terms that will be referenced here and on various websites, so they are all laid out to help avoid future confusion.

Beneficial microorganisms: Analogous term(s): indigenous microorganisms, IMO, BMO. A group of microorganisms that are cultured from the wild. Sources could include microbes found in the air, soil, or leaf mould. Specifics will be used when referring to culturing practices.

Bokashi bran: Analogous term(s): bokashi powder. A carbon substrate that is inoculated by a specific set of beneficial microorganisms. Bokashi bran can be used to inoculate organic matter in a bokashi bucket or can be added directly to the soil. The term does not refer to a specific type of bran such as wheat bran or rice bran, and can also refer to a non-bran substrate such as coffee grounds or paper. Specifics will be used when needed.

Bokashi bucket: Analogous term(s): fermentation vessel, fermentation container, bokashi bin. An airtight container used to contain organic matter and bokashi bran so they can be anaerobically fermented. The term may refer to an actual bucket, a commercially available bokashi bucket, or any other container that can be hermetically sealed and used to ferment food waste. The bokashi bucket could be homemade or purchased from a retailer. Specifics will be used when needed.

Bokashi composting: Analogous term(s): bokashi fermentation. Fermenting organic matter in an acidic anaerobic environment for a set period of time using a specific group of microbes to conduct the fermentation.

Bokashi pre-compost: Analogous term(s): fermented food waste, fermented organic waste. Organic waste that was exposed to bokashi bran and/or EM and has undergone acidic anaerobic fermentation for a period of time and is ready for final processing (finishing) — burying, mixing with carbon sources, mixing into soil, etc.

Compost pile: Analogous term(s): composting, aerobic compost, traditional composting, static pile. The common form of backyard composting involving the mixing of green and brown organic wastes. The wastes are primarily broken down in an oxidative process so the pile needs to managed aerobically in some form by aerating it and turning it or layering it to ensure aeration in a static pile. Bokashi composting will be referred to specifically as bokashi composting and/or bokashi fermentation.

Essential microorganisms (EM): Analogous term(s): effective microorganisms, EM, EM1, EM·1™, ProBio Balance Original™, ProBio Balance Plus™. A consortium of symbiotic microorganisms that are used to inoculate the organic matter and subsequently ferment it and/or used as a starter culture to make bokashi bran. EM also has a variety of other uses which won't be discussed in this text; all references will be related to its use in bokashi composting. This term will refer to the commercially available form of EM, not specific to any vendor unless explicitly noted. It will not be used when referring to wild cultured microorganisms, which will be called *beneficial microorganisms* (see definition above).

Finishing: Analogous term(s): final processing. An additional process applied to the bokashi pre-compost to allow pH to move toward neutral and assimilate with the surrounding soil, thereby making the compost safe for plants roots and/or a worm bin. This process could involve burying the pre-compost in a trench or mixing it with soil and/or a carbon source. The final product from the finishing process can be used as a part of a soil mix for planting.

Organic matter: Analogous term(s): kitchen waste, food scraps, food waste. A term used to define common kitchen wastes, including meat, dairy, bread, fruit and vegetable scraps, and coffee grounds. Any organic material that comes out of the kitchen and is used in food preparation and/or consumption. This could include raw, cooked, or processed foods, but not items such as manures, wood chips, paper, or other common compostable materials. If those types of items are discussed they will be labeled with their actual names, e.g., wood chips.

CHAPTER 1

Why Bokashi?

BOKASHI COMPOSTING AT ITS SIMPLEST IS USING MICROOR-GANISMS to anaerobically ferment organic matter in an acidic environment so it can then be rapidly assimilated into the soil by the soil biota. Bokashi composting is really a fermentation process, not a composting process. By definition, composting is an aerobic process that requires oxygen to properly compost (break down) organic matter. In traditional composting, anaerobic composting is bad and results in unfavorable by-products. With bokashi you can ferment food waste anaerobically and avoid unfavorable by-products by using a specifically selected group of microorganisms that neutralize harmful bacteria and encourage the proliferation of beneficial bacteria. Because you are fermenting organic matter anaerobically in a closed system, not composting it in the traditional sense of the word, you have numerous benefits:

- You can compost *all* types of food waste, including meat, cheese, dairy, and bread.

7

- You don't have to worry about mixing greens and browns in a specific ratio.
- No insect or rodent issues.
- No putrid odors.
- Minimal greenhouse gasses are produced.
- No loss of nutrients to the ground or the atmosphere.
- The finished product is full of beneficial microorganisms.
- Bokashi composting can be used on any scale.
- The organic waste doesn't have to be turned on a regular basis.
- Bokashi composting is much faster than traditional composting.

When you compost using bokashi you need to introduce beneficial microorganisms to the food waste to start the fermentation process. This can be done using a dry carrier such as an inoculated carbon source (e.g., wheat bran) or a liquid form via a microbial spray (e.g., activated EM). These microorganisms then go to work consuming sugars from the organic waste and the fermentation process begins. After two weeks of anaerobic fermentation, the fermented organic waste can then be applied directly to your garden soil or mixed with soil to be used as a potting mix. The simplicity of the whole process makes recycling kitchen waste very easy, and is just one of the reasons you should add bokashi to your eco tool belt.

Meat and dairy are OK

If you were to start adding meat and dairy to a traditional compost pile, you would most likely attract flies, rodents, or both. This may or may not be an issue for you and your location, but for most people it is a big issue — big enough that it can lead to neighbor complaints, stop people from composting, or even discourage them from starting.

Can you successfully compost meat and dairy in a traditional compost pile without any problems? Yes, it is definitely possible and can be done very efficiently, but it requires careful attention and a well-built pile that is actively managed. If air stops getting introduced to the pile you run the risk of the meat going anaerobic; it will smell very bad as it putrefies, releasing hydrogen sulfide and other sulfur-containing organic compounds. This is one of the reasons why even the United States Environmental Protection Agency (EPA) advises against backyard/onsite composting of these materials: "this method should not be used to compost animal products or large quantities of food scraps". [1] Having an aerobic compost pile requires a fair amount of work, dedication and knowledge. A lot of people succeed, but many come up short.

Greens and browns and the C:N ratio

Anyone who has read anything about the traditional composting process may only know one thing: you need the right mixture of browns and greens or it won't work right. This is where eyes start glazing over and the confusion begins. What is brown, what is green, how brown is something compared to something else? Potential composters start thinking and realize that they have a lot of green stuff, but where do they get all of this brown stuff? They might not have a lot of trees in their yard to supply dried leaves, and more and more people are averse to using cardboard and newspaper in their composting systems, so they are constantly in search of dried stuff to use as a carbon source. For an aerobic composting process to work optimally and quickly, you need to get the carbon-to-nitrogen ratio right, approximately 30:1. Most people don't get the ratio right, and that is why a lot of traditional compost piles fail. If you have too much brown, the pile won't heat up and fungal organisms will take over, so the pile is slow to decompose. If you have too much green, you run the risk of the pile going anaerobic and stinking (think pile full of fresh-cut grass sitting in the sun). The C:N ratio is drilled into people's heads over and over again, but it causes problems. Most people understand the concept but have problems sourcing all the right components.

When you compost using bokashi, none of that matters, because you do not need to worry about the C:N ratio. You can compost whatever you have, and what most people have are a lot of food scraps. These are ideally suited to be composted using bokashi. They are usually sized down already, and most are easy to ferment in a bokashi system because they contain a lot of sugars. You just need to collect the food scraps, add them to the bucket, inoculate them with bokashi bran, and wait. Once the food scraps are introduced into a bokashi system, the microorganisms will start fermenting the scraps immediately, the pH will drop, and no putrid odors will be generated.

No pest problems or putrid odors

When you compost meat, dairy, or any other food scraps with bokashi, you don't have to worry about putrid odors or attracting pests. All bokashi systems ferment the food waste anaerobically in a sealed container. So no smells can emanate from the food waste to attract pests. Even if ever-curious flies, rodents, or pets are in the area, they can't get at the food waste because it is sealed inside the bucket and they are stuck outside.

The fermentation process itself should not produce any putrid odors, though it will create a smell that the bucket will keep contained. Even though the fermentation process suppresses putrefaction and the rancid odors that accompany it, the process does produce a typically vinegar, pickle smell that is closer to pleasant that putrid. But again, the container is sealed shut so you can ferment your food waste inside your home or apartment and even the most sensitive noses won't be able to detect an odor. This is a huge advantage of using a closed system — the contents stays sealed away inside of a container while the microorganisms ferment the waste.

In addition to not giving off offensive odors, the microbes within the bokashi bran suppress putrefaction, pathogenic and methane-producing microbes, dramatically reducing the greenhouse gasses that are produced.

Minimal (if any) greenhouse gasses are produced

One of the dirty little secrets in the composting world is that traditional composting methods inherently generate greenhouse gasses (GHG). This is never mentioned in traditional composting circles, but is actually a pretty big problem. Methane, carbon dioxide, and nitrous oxide are all by-products of the traditional composting process, and all three are greenhouse gasses.

Aerated composting (AC), one of the most common routes for recycling organic carbon into soil, has undeniable economical and ecological advantages, but also has some notable shortcomings. Its carbon recycling potential is 50 percent or less, the entire process is long (≥ 6 months), and its GHG footprint is very large. The main gas produced during AC is CO_2, but CH_4 (~25 times more potent as a GHG than CO_2) is also a notable by-product. Along with C mineralization, N is also released, mostly as amines, heterocyclic compounds, ammonia, nitrite, and nitrate. Suboxic, acidic, and organic-rich conditions can lead to incomplete denitrification with the formation of N_2O as well. The amount of N_2O emitted is small relative to CO_2, yet N_2O is ~300 times more powerful as a GHG than CO_2. There is furthermore clear evidence that unchecked N_2O produced during the turnover of carbon residues amended in soil can contribute GHG equivalents to the atmosphere which more than offset carbon savings due to aerated composting.

— Green and Popa,
"Turnover of Carbohydrate-Rich Vegetal Matter"

So traditional (aerated) composting isn't as much of a free lunch as some would have you believe. When you compost organic matter aerobically using traditional methods, you will generate a high level of greenhouses gasses like methane and nitrous oxide, in addition to a lot of CO_2. That is why the compost pile gets measurably smaller — a lot of the carbon is volatized off into the atmosphere as the organic matter breaks down. For an eco-conscious person, that isn't good. It is MUCH better to have that carbon tied up in the soil matrix than it is to have it in the atmosphere.

Even under the best aerobic composting conditions, greenhouse gasses are emitted. But in reality most people don't maintain perfect aerobic compost piles; they let them go anaerobic, and that is a cause for concern. How many people start actively aerating a compost pile but then give up over time, leaving it untouched and anaerobic? A lot. When organic material composts anaerobically, nitrous oxide, ammonia, and hydrogen sulfide are produced. That is why anaerobic compost piles stink; the smell is a combination of these gasses. The ammonia not only smells but also leaches out of the pile into the ground, where it can potentially contaminate ground water. This also takes the nitrogen out of the compost, which is not what you want — you want nitrogen in your compost, where it will ultimately feed your plants. Methane is also produced in the pile when methane-producing microbes take over. These microbes prefer anaerobic conditions at a neutral pH; under those conditions, they can dominate the pile and multiply rapidly, producing a lot of methane. This is how they produce methane in bio-digesters, where it is harvested and used for fuel in industrial situations. The average home composter isn't harvesting the methane so the gas goes into the atmosphere. All of this GHG production can be avoided by fermenting food waste using bokashi instead of breaking it down in an oxidative process.

When you compost with bokashi, you are fermenting organic waste anaerobically at a low pH so greenhouse gas production is drastically reduced. Methane-producing microbes can't survive at a low pH, so very little methane gas is produced. But what about the other microbes involved — don't they generate gasses? Lactic acid bacteria are one of the primary constituents of the group of microorganisms used in bokashi bran to ferment food waste, and lactic acid fermentation is a process that does not generate any gasses. In a study, Dr. Lawrence Green found that:

> Organic waste processed by bokashi fermentation produces no measurable gas during the 7-day fermentation process and when then mixed with soil it is further degraded without evidence of any gasses being liberated. Based on these findings it appears that bokashi fermentation does not produce measurable gas emissions in its conversion of organic waste into a nutrient-rich end product that can be used to support plants and crops.
>
> — Green, "A Pilot Study Comparing Gaseous Emissions"

Unscientific visual observation shows that no excess gas is produced when you compost using bokashi in an anaerobic container; the container doesn't swell, and there is no audible *poof* when it is opened after a two-week fermentation. This is clearly an area where more research is needed, but based on the science of the processes involved and the information available today, it seems that few if any greenhouse gasses are produced during the bokashi fermentation process.

No loss of nutrients — everything ends up where you want it, in the soil

Another benefit of fermenting organic matter inside a closed container is that all of the nutrients are retained in the process.

Nothing is lost through runoff or ground penetration. None of the nitrogen is lost to ammonia, and none of the carbon is oxidized off into the atmosphere. All of the amino acids, vitamins, enzymes, and nutrients generated or liberated from the organic wastes in the fermentation process stay contained within the bokashi bucket and ultimately end up exactly where you want them, in the soil.

A lot of carbon is lost in the traditional composting process when it is volatized off as carbon dioxide. In addition to being a greenhouse gas, it has been argued that this carbon is needed when the organic matter is added to the soil in order to ultimately produce polysaccharides, which help improve the soil structure. By fermenting waste in a closed container without producing CO_2, you are putting the carbon and other organic materials directly into the soil when you use the bokashi pre-compost. This may arguably benefit the soil structure more than the traditional method of applying already composted organic material, which is devoid of a lot of the original carbon.

The closed container also eliminates any evaporation, and thus the need to add additional water to the system. With water scarcity becoming more and more of an issue worldwide, this is a big advantage. The fermenting contents stay moist throughout the process, and the finished product is still wet when it is applied to the soil. This initially adds moisture to the soil where the bokashi pre-compost is applied while increasing the long-term water-holding capacity of the soil by supply organic matter that can be converted into humus by the soil biota. A lot of other soil amendments and composts are applied to the soil either dry or on the drier side, so they can act as a wick and draw moisture out of the surrounding soil during their initial application.

The final product is full of beneficial microbes

The microbes inoculated on the bokashi bran and used in the fermentation process aren't just valuable for fermenting food waste;

they are also valuable when they are introduced to the soil. The main microbes within the bokashi bran include lactic acid bacteria (LAB), yeast, and purple non-sulfur bacteria (PNSB). This group of microbes has the ability to perform a variety of beneficial functions including the breaking down of harmful chemicals and wastes and the ability to create bioactive substance and beneficial enzymes. All of the microbes in EM·1™ (the most common inoculant for bokashi bran) are derived from nature and occur naturally in soils and waterways around the world. These microbes were specifically selected by Dr. Teruo Higa for their ability to perform their own unique individual functions and function cooperatively in a consortium of beneficial microbes. The consortium of microorganisms have a lot of uses and real-world applications: they are used in sewage treatment plants because they have the ability to break down wastes and pollutants; on farms to help control manure odors; and to process food waste, which they are able to ferment, thereby speeding up its breakdown.

When you inoculate kitchen waste with bokashi bran, you are introducing these essential microorganisms to a food source. The microbes then start eating the sugars in the waste and begin to multiply and ferment the food waste. Some organisms also begin to feed on the waste of other organisms in the system. For example, the PNSB will feed on dead yeast and LAB bodies, and the LAB will consume some of the waste products of the PNSB.

Fermentation greatly increases the populations of these microbes. After two weeks of anaerobic fermentation, the process just doesn't suddenly stop because we said the time is up; the microbes continue to act on food waste. When you then combine this fermented food waste with soil, you introduce the microbes into the soil. Some of these microbes ultimately survive and multiply in the soil, where they have a beneficial influence, mainly through their ability to break down wastes and suppress pathogens, thereby improving soil quality. Introducing uninoculated food waste to the soil won't have this same effect. In addition, raw food waste will release greenhouse gasses when it starts breaking down in the soil. Dr. Teruo Higa, the creator of EM, has stated that "kitchen refuse decomposes if simply buried in the soil, giving off offensive odors, evidence that degenerative microorganisms are at work. EM added to matter in this condition brings about a change in the microbiological equilibrium, causing dominance to shift away from degenerative to regenerative strains of microorganism. Under these conditions, organic matter no longer gives off offensive odors."[2]

Full scaleability: you can ferment a little or a lot

Bokashi composting can be scaled up or down in size depending on the needs of the end user without requiring any extra work or processing. In contrast, the effectiveness and work required for a traditional aerobic compost pile will vary with size.

A traditional compost pile needs to be about one cubic meter in size to work effectively. If it is any smaller than that, the pile won't build up enough heat to break down properly, and anything larger than one cubic meter will generate a lot of heat, requiring different, more energy-intensive management strategies. So traditional compost piles can be scaled up in size, but the tradeoff is more labor and/or more time required to get to a finished product. And they can't really be scaled down much smaller than a cubic meter without requiring more monitoring and time.

With bokashi composting, size is not an issue. The amount of mass being fermented has no effect on the final product; large amounts will ferment the same as small amounts. All bokashi systems need to be able to do two principal things, regardless of size: drain excess fluids from the system and maintain anaerobic conditions within the system. If both those conditions are met, it is possible to ferment one pound or ten thousand pounds of waste using the same process. As long as the environment is maintained, then the steps are the exact same, regardless of size. No extra energy, equipment, or water is required to compensate for increased size.

No turning required and no extra water needed

In traditional composting, organic matter is broken down in an oxidation process, so you periodically have to turn the pile to introduce oxygen to other parts. Turning also moves the contents on the outside of the pile to the inside, where they can heat up and break down. Now, could you manage that same pile without turning it? Sure — but not without increasing the amount of time it will take to obtain finished product. A static, semi-aerobic compost pile may take up to a year to fully break down into finished compost. It also runs the risk of going to the bad side of anaerobic, because air isn't being constantly introduced to the center of the pile, and this could ultimately create odors.

Since traditional compost piles are usually exposed to the open air, you may have to add extra water to them throughout the process to maintain the proper moisture level, so they heat up and the contents are broken down properly. Some of that water will drain right through the pile into the ground, some will evaporate, and some more will be absorbed by the pile and used by the microorganisms that break down the organic waste. Regardless, additional water is required, some of which will leave the system in one form or another, so it has to be replaced somehow.

With bokashi, the system is closed, and the organic matter you add is generally at least 30 percent moisture, so you don't need to add any extra water regardless of your system's size. Some water will be removed when the bokashi leachate is drained from the vessel, but not enough to require the addition of any extra water.

Since the fermentation needs to take place anaerobically you do not want to add extra oxygen into the system by opening the vessel and turning the contents. This saves you a lot of work compared to traditional composting. Once you have completely filled the system, all you have to do is drain off the leachate and wait for the fermentation process to run its course. And since the process is standardized regardless of size, you can ferment your food waste in any size vessel. This is a big advantage for people with limited space. People with small apartments, offices, or yards can compost their food waste in a corner of their garage, closet, or anyplace else, in spaces where they couldn't fit a traditional compost pile.

Bokashi composting is fast

Bokashi composting is much faster than traditional composting. With bokashi you can turn raw kitchen scraps into something you can plant into in 30 days. During those 30 days, you are just waiting most of the time, as the process requires little work.

In contrast, a well-built traditional compost pile can only turn organic matter into humus in about 30 days if it is extremely well made and actively managed. The average person won't be able to accomplish this for a number of reasons, the main ones being that you really have to know what you are doing and be willing to put a lot of work into the pile. I know from personal experience that most people don't want to go out and turn a compost pile every other day just to maximize efficiency. They just won't do it, it is too much work. So from a work requirement standpoint the 30-day goal of traditional aerobic composting makes it a bit of a long shot. Another key requirement to get good-quality compost in 30 days from a traditional pile is quantity and quality of ingredients. You need to have a cubic meter worth of compostables, and they need to be in the ideal 30:1 ratio of browns to greens; again, it is a difficult task for most people to achieve this on a consistent basis. Most people generate the bulk of their waste organic matter in the kitchen, and you can't just pile up kitchen waste and compost it without running into the problems discussed earlier.

Bokashi is a logical fit for this niche. You can add straight kitchen waste to the fermentation vessel as it is generated, apply some bokashi bran, walk away, and in 15–30 days you will have a finished product that you can put straight into the ground. Two weeks after that and you can plant directly into the finished bokashi compost. That is fast in compost timeline terms, especially considering there is little work and no mixing of ingredients required.

Bokashi gives you finished compost in 30–45 days all in, maybe even faster with some extra effort. The speed of the fermentation process is a big advantage. It also allows you to cycle organic waste faster by going straight from kitchen to fermentation to soil over and over again in small batches. You don't need to collect a critical mass of scraps to start with, so the fermentation process can start right away. You can also scale the fermentation vessel to the amount of scraps you are generating, so they go straight into the bokashi bucket as soon as you produce them.

Compared to traditional composting, this is a big advantage. You don't have to worry about what to do with kitchen scraps while you collect all of your other materials and amass enough volume to get started with your one-meter pile. I have heard of people freezing kitchen scraps until they have enough materials to assemble a full-size compost pile. That is a lot of extra work and room taken up in your freezer, and if you step back and think about it, it is kind of ridiculous. Feed the scraps to chickens or worms or compost them some other way like bokashi, but don't freeze them. Why go through all that trouble? It just isn't practical for everyone.

So now you have to add material collection time to the 30 days it takes to compost in the traditional method (at its fastest). Depending on the situation, that could be days or even weeks, so the 30-day time frame is out the window given that most people won't just have all of their scraps given to them on Day

One. And a lot of people will never generate enough scraps in a timely manner to build a one-cubic-meter compost pile, so they need a system that can break down food waste as it is generated, like bokashi. When you start comparing timelines, the amount of available organic waste, and the amount of work required, you can start to see that bokashi is more practical than traditional composting for homeowners who only deal with kitchen waste on a regular basis.

Conclusion

All in, bokashi composting requires less work than traditional aerobic composting, is easier, and may be more practical for some people looking to recycle their kitchen waste. Bokashi composting fills a niche. It takes away a lot of the reasons why the average person won't, or doesn't, compost their kitchen waste. You don't have to worry about moisture content, you don't have to worry about the C:N ratio, you don't have to spend time aerating the pile, and you can compost small (and large) volumes of food waste in a small space like an apartment. All of these factors can be difficult to manage and require a considerable amount of time and energy, sometimes more than the average person wants to spend. Even if you do all of these things you might end up waiting a month or more only to discover that you really don't have that great of a finished product anyway. Then most people give up and move on, saying, "Composting doesn't work for me."

Bokashi fills that niche and makes things easier and more predictable. Fermenting food waste with bokashi is faster, so you can cycle nutrients through the system more rapidly with more consistent results. Sure, you have to buy the inoculant or make it yourself, but it is relatively cheap and easy when you factor in the benefits. And the inoculant (and microbes) ultimately end up in the soil because everything is fermented in a closed system with no losses. The process yields a nutrient-dense, microbe-rich finished product that is easily broken down by the soil biota. This finished product, called bokashi pre-compost, contains microorganisms that have the ability to suppress and outcompete pathogens and disease organisms in the soil. It also builds soil tilth and increases water-holding capacity by adding organic matter to the soil. And most importantly, it gives an eco-conscious person another tool for recycling their waste into an Earth-friendly finished product.

CHAPTER 2

The History of Bokashi

THE VERY WORD *BOKASHI* INVOKES A SENSE OF MYSTERY. The exact translation from Japanese adds to the enigma. I have come across various translations, with the most common being "fermented organic matter." I have also seen *bokashi* translated as "obscuring the direct effectiveness." The first translation makes sense because it directly describes the process involved. The latter is a bit more confusing — to obscure the direct effectiveness, to conceal it, to make it difficult to understand. On the surface, that definition makes some sense, because the processes involved in bokashi are a bit difficult to fully understand and the whole process is often downplayed as a form of voodoo composting. But I think that the concept of "obscuring direct effectiveness" stems from the ability of the bokashi fermentation process to alter the original materials enough to make them unattractive to pests, yet still result in an end product that is highly beneficial for soils and plants. Michinori Nishio has written that "if rape seed or soybean meal is directly applied to soil, a certain fly lays eggs in it. The maggots feed on young seedlings and cause serious damage. Fishmeal also attracts field mice, which dig tunnels under seed beds. To avoid damage of this kind, farmers developed on their own initiative a technique of composting organic fertilizers for a short period — bokashi."[3] He then went on to say that "overall, the aim of the process seems to be, firstly, to decompose substances which attract pests, and secondly, to create a slower-acting organic fertilizer."[4] These two statements put the second translation into perspective. As is often the case, the problem lies in the translation. Bokashi's roots lie in Asia, and a lot of the research and information about it is written in Japanese, which literally leaves some holes in the translation of information.

Today bokashi composting systems are used worldwide as an alternative to traditional composting. Although the exact path that bokashi has taken to get to the inoculated bagged bran and plastic bucket we know today isn't obvious, it is clear that modern-day bokashi has its origins in the Far East. It was there that relatively recent scientific research merged with traditional farming methods, resulting in a fermented mixture of beneficial microorganisms and a carrier: what we know as modern-day bokashi.

The roots of bokashi are wrapped around the traditional natural farming philosophy practiced in Korea and other parts of Asia. One aspect of that philosophy that is related to modern-day bokashi is the culturing of indigenous microorganisms, or IM; these are naturally, locally occurring microorganisms that are cultured onto a substrate such as cooked rice or milk. Many of the "wild" cultured microorganisms are the same organisms found in most modern bokashi bran that has been made using essential microorganisms (EM). In natural farming, the IM are used to inoculate compost piles to make semi-aerobic bokashi compost, because they provide a beneficial microbial source to help break down the organic matter quicker.

In the last century, the natural farming movement was advanced by people such as Mokichi Okada; today its leading advocates include Dr. Han Kyu Cho. The tie to Japan lies in the discovery and formulation of EM by Dr. Teruo Higa in the late 1970s and early '80s. Dr. Higa was born in Okinawa, Japan, in 1941. As a young boy, he worked in agriculture and was well acquainted with the hard labour that went along with it, including making compost, which he hated. But he had a passion for growing food that he went on to pursue at the University of the Ryukyus, the most prestigious university in Okinawa. After graduating from the agriculture department there, he went on to get a doctorate in agriculture research from Kyushu University in Fukuoka. He then returned to Ryukyus and began teaching, eventually becoming a professor in 1982. There he spent his time focusing on the cultivation of mandarin oranges, a large part of which involved the use of chemical fertilizers, which he supported ardently at the time. His research meant that he spent a lot of time in the field, oftentimes in contact with the chemicals he believed in. But the exposure to the chemicals eventually led to health problems, and he began to wonder if there was a better way.

One day, while conducting some research on watermelons in the Middle East, Dr. Higa had an epiphany — "he came to realize that agriculture had come to rely far too heavily on the use of chemicals, and decided to find a better approach where something like microorganisms could be used to manage plant growth."[5] He was aware of existing research into agricultural microorganisms and began conducting his own studies, looking for microorganisms that could universally support plant growth.

One day during this research, Dr. Higa discarded a combination of test microorganisms from some research work onto some grass near his laboratory. He paid no attention to it at the time, but as time passed, he noticed that the grass that had come into contact with the waste mixture of microorganisms was extremely healthy and vibrant compared to the other grass nearby. He came to the conclusion that the microbes had contributed to the health of the grass and, most importantly, that the mixture included all the microbes he was researching at the time. It was the combination of those microbes that had such a beneficial effect on the plant growth; it was all of the microbes acting as a consortium, not any one microbe acting alone.

This seemed impossible and was contrary to all his research; the general thinking of the time said that combining microbes should result in antagonistic behavior, with every type of microorganism competing negatively with other types. Nevertheless, Dr. Higa used this accidental discovery as a stepping stone and continued his research by collecting and mixing a large number of microbes from all over the world in his search for the optimal blend. He continued to fine-tune the mixture until the late 1970s, when he finally developed EM1, or effective microorganisms. This combination of microbes fulfilled all of his criteria as a replacement for synthetic chemicals, "the promotion and maintenance of healthy plant growth resulting in more abundant harvest of better tasting crops".[6]

EM1 is an optimized blend composed primarily of lactic acid bacteria, yeast, and phototrophic bacteria that work together synergistically as a group, creating a system for mutual support. The member microbes in the consortium eat the waste of other microbes, while simultaneously synthesizing beneficial by-products such as enzymes, anti-oxidants, and vitamins that the other microbes can use. The EM1 consortium is dominant over other microbes in nature and has the ability to reeducate and outcompete harmful pathogenic bacteria, thereby reducing odors and disease. These properties led the way for EM1 usage to spread into the waste management and animal husbandry sectors, where it is used to help deodorize and clean effluent. EM1 was first marketed in Japan for a wide variety of uses in 1982 and is now used in over 120 countries worldwide.

So where did EM1 and natural farming's bokashi composting meet to become the fermentation process EM bokashi? The most likely tie between the two is Dr. Higa's relationship to Sekai Kyusei Kyo (SKK), a religious organization in Japan that promoted a type of natural farming known as Kyusei Natural Farming. Sometime after Dr. Higa's discovery and formulation of EM1, the two parties connected. Dr. Higa is not now and never was a follower of Sekai Kyusei Kyo, but his goals for his EM1 mixture and those of the practitioners of nature farming within SKK were aligned, in that both wanted to move away from agricultural chemicals to allow food to be safer, more nutritious, and more sustainable. The new EM1 product gave natural farming practitioners a way to achieve this goal, so there was an inherent synergy there.

One component of natural farming was bokashi, and farmers across Asia had practiced various forms of bokashi fermentation for generations using indigenous microorganisms. The introduction of EM1 as a microbial source for nature farmers seemed like a natural fit. The microbes in EM1 were optimized and cultured in a laboratory so the mixture was consistent in quality. The product was also readily available because it could be produced in large quantities and distributed easily. Dr. Higa mentions the use of EM1 in natural farming applications in his 1993 book *A Earth Saving Revolution:* "Amazing harvests are being obtained using absolutely none of these substances [artificial chemicals and fertilizers] at all but a fermented mixture of rice bran, rice straw, fish meal and EM concentrate called 'EM Bokashi' spread over the area to be cultivated together with an appropriate amount of liquid EM in place of regular compost. The result: larger harvest of better quality produce which is complete free from harmful chemical residues".[7] The growing availability and convenience of EM1 now made it possible for more people to start practicing bokashi composting.

Today people all over the world use EM1 to make semi-aerobic bokashi compost and the more familiar anaerobic fermentive bokashi compost. In the latter Dr. Higa's EM1, or a similar product, is used as the main microbial source in the fermentation process. It is this specific consortium of microbes that gives bokashi composting some amazing properties.

CHAPTER 3

The Science

WHENEVER SOMEONE MENTIONS *anaerobic* and *composting* in the same sentence, people knowledgeable about compost are trained to immediately say, "Oh no, no, no, that's bad." This stereotype is a half-truth. Allowing organic matter to go anaerobic and putrefy in an uncontrolled environment is absolutely bad and should be avoided at all times. But allowing organic matter to *ferment* anaerobically is perfectly acceptable. It is acceptable because in the latter case we are controlling the process using microbes that we have specifically selected in order to obtain a desired result. When we anaerobically ferment food waste with bokashi, we are providing the conditions for the beneficial bacteria to dominate and outcompete the putrefactive anaerobic bacteria, thereby removing the reasons for the stereotype — slimes, gasses, and smells.

"Composting" with bokashi is actually a fermentative process, not the oxidative process traditionally associated with composting. Dr. Steve Diver describes the fermentation process as "the production of useful substances (alcohol, amino acids, organic

acids, and anti-oxidation substances) via metabolism of microorganisms (beneficial microorganisms)."[8] It is these by-products
that add to the nutrient value of the finished product. To make
sure that the by-products are good rather than harmful, we delib-
erately inoculate organic waste with a specific group of microbes,
ensuring that it is fermented, not putrefied (putrefaction is the
production of harmful substances). During this fermentation
process, the microbes produce by-products that are able to suppress the microbes that cause putrefaction and disease:

> For example, part of the fermentation process involves the
> production of organic acids such as acetic acid, butyric acid,
> propionic, and lactic acid, amongst others. As the acids are
> produced, the pH drops, decomposition stops and the fer-
> mentation process begins. The low pH environment has the
> ability to inhibit and destroy pathogens. For example, the
> beneficial effects of two of these acids were studied in Korea
> where they found that acetic acid and butyric acid were
> effective soil fungicides and could potential help in the con-
> trol of soil pathogens such as fusarium and phytopthora.
>
> — Kim, Hong-Lim, Bong-Nam Jung, and Bo-Kyoon Sohn,
> "Production of Weak Acid by Anaerobic Fermentation of Soil
> and Antifungal Effect"

A flow chart made by Dr. Teruo Higa and Dr. James Parr illus-
trating the advantages of fermentative decomposition of wastes
is shown on page 31.

The microorganisms that ferment the organic waste are the
single most important part of bokashi fermentation. There are
various ways in which they can be introduced into the process,
but the most common way is via a carrier inoculated with EM.
Bokashi fermentation conducted with EM is typically known as
EM bokashi, specifically denoting that EM (not IMO, lactobacillus serum, or something else) is the microbial source.

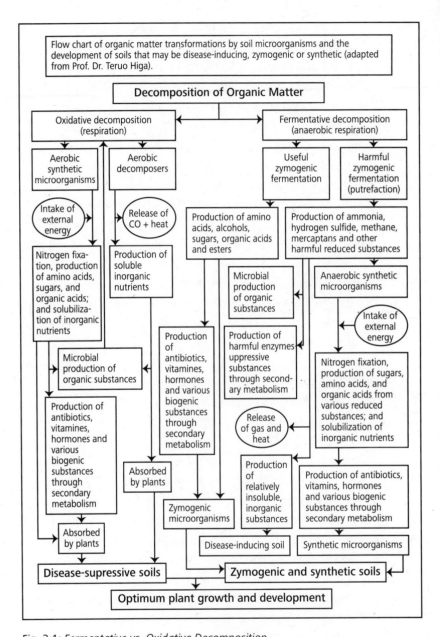

Fig. 3.1: *Fermentative vs. Oxidative Decomposition*

SOURCE: *FOR A SUSTAINABLE AGRICULTURE AND ENVIRONMENT*, HIGA AND PARR, 1994

EM is a synergistic consortium of microorganisms, a group of more than one type of microbes that works together so that each organism causes benefit, not harm, to the other organisms in the consortium. As described in Chapter 2: The History of Bokashi, the original EM product, EM1, was formulated and popularized by the Japanese researcher Teruo Higa, who spent many years trying to get individually beneficial microbes to function together in a consortium which would then be even more beneficial. He knew it was important to develop a consortium because individually many of the microbes that he saw as beneficial only functioned under a very specific set of environmental conditions which made their use very limited or impractical in widespread, real-world applications where conditions can vary greatly. By creating a successful consortium, he was able to greatly increase the ability of the member microbes to function over a larger range of temperature, pH, and environmental conditions. This adaptability and versatility made the end product much more stable and usable in the real world. EM1 was universally applicable, something that Dr. Higa had striven for in his quest to find a microbial product to ultimately replace harmful agricultural chemicals. In *An Earth Saving Revolution*, Higa describes EM as:

> a large group of microorganisms…. photosynthetic bacteria, yeasts, lactic acid bacteria, and fungi are just some of the strains of anabiotic microorganisms belonging to the EM group. When a combination of them is present in the soil and they are proliferating in sufficient numbers they bring about an increase in antioxidation levels and a resultant intensification in energy concentrations. In other words, their activity instigates the regeneration process, purifying the air and water content of the soil and intensifying plant growth. A further positive feature of anabiotic microorganisms is that their secretions contain

large amounts of nutrients beneficial to both plants and animals including amino acids, organic acids, polysaccharides, and vitamins.[9]

All of the microbes in EM1 and EM1-type products are naturally occurring microbes from around the world. None of them are genetically modified in any way. At a minimum, all quality EM cultures will contain various species of yeasts, lactic acid bacteria, and purple non-sulfur bacteria (PNSB). The total number of species of each type of microbe in any given bottle of EM mother culture can vary from a few to more than one hundred. Each manufacturer formulates their products slightly differently, based on their research and the intended end use for the particular product. For example, I have one bottle of an EM-type product that states that it includes at least seven species of microbes (probably many more are included, but not listed on the bottle):

- *Lactobacillus plantarum*
- *Lactobacillus casei*
- *Lactobacillus fermentum*
- *Lactobacillus delbrueckii*
- *Bacillus subtilis*
- *Saccharomyces cerevisiae*
- *Rhodopseudomonas palustris*

Another bottle of a very similar product made by a different manufacturer states that it contains these 14 species (and again, probably many more not listed):

- *Bacillus subtilis*
- *Bifidobacterium animalis*
- *Bifidobacterium bifidum*
- *Bifidobacterium longum*

- *Lactobacillus acidophilus*
- *Lactobacillus casei*
- *Lactobacillus delbrueckii subsp. bulgaricus*
- *Lactobacillus fermentum*
- *Lactobacillus plantarum*
- *Lactococcus lactis subsp. lactis*
- *Rhodopseudomonas palustris*
- *Rhodopseudomonas sphaeroides*
- *Saccharomyces cerevisiae*
- *Streptococcus thermophilus*

So here are two commercially made, high-quality EM1-type mother cultures, and each has a stated different number of active microbial species — different manufacturers, different formulations. Without doing any in-depth laboratory analysis, both products perform equally well in terms of making a high-quality bokashi bran; I cannot notice any difference between the two. It is my belief that, as long as you are buying an EM-type product from a quality manufacture, the exact species makeup of the EM formulation isn't important, because the main species will always be present and will drive the consortium. The other support species will add to the depth of the consortium with the functions that each performs. Dr. Higa has always stressed that it isn't the exact combination or ratio of microorganisms that makes EM so powerful, it is the fact that the microbes are working together as a group, supporting each other and feeding off of each other, so no one microbe becomes too important, making the group itself much stronger and more adaptable. This synergy and diversity give the consortium the ability to persevere and adapt to a wider array of environmental conditions, as well as giving microbes that would be unable to survive individually the ability to be used successfully in consortium. This attests to the power of having these microbes operating and acting in a consortium rather than as individuals.

As a consumer of EM, you will see that formulations can vary from place to place, and may very likely change from bottle to bottle. Yet high-quality EM mother culture still remains effective. I feel that over time you will realize that if you buy a high-quality product, then you will get high-quality results, regardless of minor formulation variances. So, like anything else, it is important to always buy a high-quality product from a reputable manufacturer. In the US, there are two main sources for EM1-type mother cultures:

TeraGanix (teraganix.com) that distributes EM1 for EMRO USA.
SCD Probiotics (scdprobiotics.com) that manufactures ProBio Balance.

Both products are sold at arm's length by the manufacturers themselves and through other retailers who have resale agreement

Fig. 3.2: *TeraGanix EM1 (right) and SCD Probiotics ProBio Balance Plus (left).*

with the manufacturers. Both are very high quality and work tremendously well. I have no problem highly recommending either one.

For worldwide sources of EM1, visit emrojapan.com.

Regardless of anything that is written out there on the Internet, it is not possible to recreate or manufacture EM in a home-based setting. The techniques and equipment needed to get the microbes in the EM mother culture to function in consortium are far above and beyond what any normal homeowner would have or could do. Homebrewers can kind of make EM-like cultures, but they won't be able to recreate anything with the lifespan, depth, or diversity of an EM mother culture. So ignore what you read — you can't make EM at home. Most homemade EM recipes are just lactobacillus serums — effective for some things, but not EM, which contains a variety of microbes, not just lactobacillus. Later I will delve into a microbial solution that a homebrewer can make, but again, it won't be able to compete with EM in terms of shelf life, effectiveness, quality, or ease of use — and EM is cheap when you get right down to it. Most people would be better off purchasing a high-quality EM mother culture when they make bokashi bran because it gives them a known ingredient versus an unknown variable (homemade culture). That means a more consistent product, and less time wasted trying to get the process to work the way that you want it to work.

So why is EM so important to the bokashi composting process? In its simplest form, bokashi composting is a fermentative process, and for the fermentation to successfully happen, the microbes being applied to the food waste need to have the robustness to outcompete and control any existing microbes that are already in the system. EM does this. Thanks to the consortium of microbes within EM, bokashi bran inoculated with it has the ability to reeducate other "wild" microbes. Vinny Pinto, a longtime researcher in the area of syntropic antioxidative microbes and EM, describes the primary species in EM as "dominant organisms, meaning that they can entrain and control other local environmental organisms, which are more passive and often open to entrainment".[10] This is why you shouldn't add a lot of already spoiled or moldy food into your bokashi bucket. The fermentation process works on this domination principle, and if there are too many "bad" species of microbes present at the beginning, the microbes in the EM bokashi bran may not be able to dominate and control them unless you add large amounts of EM, either through liquid spray or extra EM bokashi bran. It also explains why you should add more bokashi bran when you start detecting problems with your bokashi bucket, or when you add harder-to-break-down items such as meats to it. More EM bokashi bran means more good microbes, and more good microbes have a better chance of dominating and entraining bad microbes. When the bad microbes are suppressed and controlled, you will get what you are ultimately looking for: a fermented food waste that is alive with beneficial microbes.

Vinny Pinto describes some of the ways that EM organisms are able to alter their environment to discourage the growth of harmful microorganisms, such as:

- when in anaerobic fermentation, the EM organisms produce copious quantities of lactic acid and smaller amounts of malic, acetic, propionic, and benzoic acids. These serve to lower the pH, often to below 4.0, which strongly discourages the growth of many so-called harmful organisms.
- many of the above-named acids are mildly antioxidative, and this antioxidant action will also discourage the growth of many undesirable organisms.
- when in anaerobic fermentation, the EM organisms produce a number of anti-microbial substances which tend to destroy or discourage a number of undesirable organisms.[11]

This information is echoed by EMRO, the EM Research Organization, which states that:

> EM has the potential, given the conditions, to suppress the putrefactive microorganisms and dominate this sphere and creates re-animated surroundings, that is, organics are transformed through the process of fermentation as opposed to putrefaction, and living organisms, as well as inorganic materials, are enabled with the means to impede deterioration.[12]

The ability of the microbes in the EM consortium to produce antioxidants and suppress pathogens are just a couple of the reasons EM has gained a lot of attention by researchers. Let's now take a look into EM and see what the individual microbes are that make the culture so special.

EM (effective microorganisms)

EM is primarily made up of three groups of microorganisms: lactic acid bacteria, photosynthetic bacteria (also know as purple non-sulfur bacteria or PNSB), and yeasts. As previously discussed, each EM formulation may contain a variety of species from each of the groups. EM mother cultures may also contain lower concentrations of other microorganisms such as actinomycetes and fungi, but the bulk of the microbes in the culture are the big three:

Lactic acid bacteria (LAB)

Lactic acid bacteria is, taxonomically, a generic term for bacteria that convert large amounts of sugars into lactic acid through lactic acid fermentation. Through the production of lactic acid, lactic acid bacteria also inhibit the growth of pathogenic microorganisms and other various microorganisms by lowering the pH. Lactic acid bacteria are widely known in the production of fermented foods such as cheese and yogurt that can be naturally preserved for a long period of time.

— EMRO Japan, "Microorganisms in EM"

LAB provide multiple functions to the microbial consortium. They drop the pH of the fermenting contents down into the mid to low 3s (acidic). This both creates an environment that pathogens and methane-producing bacteria can't survive in and preserves the nutrients within the food waste, preventing them from being volatized into the atmosphere. Vinny Pinto describes the lactic acid bacteria as providing an environment for the other microbes to live in.

In this classical "nucleus/orbital" atomic model, the lactic acid bacteria are seen as providing literally a safe environment or "housing" for the other two groups of organisms,

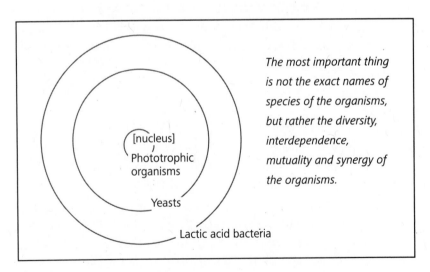

Fig. 3.3: Source: Pinto, "Introduction to Effective Microorganisms (EM)"

as do the yeasts to some extent as well, so each of these is shown as occupying a "shell" around the nucleus, which is the phototrophic organisms. The phototrophic bacteria are seen as rather dependent upon the two types of organisms in the shells to provide the necessary conditions to survive and thrive, but, on the other hand, they also provide vital nutrients to the organism in these shells. The reason the phototrophics occupy a position in the center is not only because they are protected and nurtured by the shell organisms, but because they are the "heart" of much of the magic of EM: they provide a reducing (meaning electron-donor, versus oxidative) environment, replete with readily-available hydrogen in the form of dissolved gases and various hydrogen ions (including hydride species), and marked by a relatively low ORP (oxidation-reduction potential), which, along with other secretions of the phototrophics, enable the trio of groups to decompose organic materials in a reducing manner, rather than an oxidizing manner. The three types of organisms work together to help each other digest a wide range of materials and to produce a wide range of antioxidants and nutrients.

— Pinto, "Introduction to Effective Microorganisms (EM)"

The consortium of microbes in EM is very powerful, and lactic acid bacteria play a key role. Crucially, without them there would not be any fermentation. A homebrewer can take advantage of the beneficial properties of lactic acid bacteria by culturing their own LAB serum (more on this later). Such homemade serums are pure lactic acid bacteria, so they lack the benefits the other microbes in the consortium provide, but they can provide some uses on its own. But in the EM consortium, LAB are just one of the three main components, yeast being another one.

Yeast

Yeasts are single-cell fungi that are found everywhere; they are in the air, the soil, on plants, and on you. Yeasts decompose sugars such as monosaccharides and polysaccharides by secreting enzymes that break down the organic matter into a form that the yeast cells can then absorb back into their bodies through the cells walls. During the decomposition process, yeasts produce various beneficial by-products such as vitamins, hormones, and amino acids.

> Yeast synthesize antimicrobial and other useful substances required for plant growth from amino acids and sugars secreted by phototrophic bacteria, organic matter and plant roots. The bioactive substances such as hormones and enzymes produced by yeasts promote active cell and root division. These secretions are also useful substrates for effective microbes such as lactic acid bacteria and actinomyces.
>
> — SCD Probiotics, "SCD Probiotics Inside"

Commercially available compost starters or compost activators often contain yeast in the form of *Saccharomyces cerevisiae*.

Purple non-sulfur bacteria, PNSB:

Purple non-sulfur bacteria are a group of bacteria that are able to survive under a variety of different conditions and adapt their energy production to the conditions at hand. For example, when light is present, they are able to use photosynthesis to derive energy, and when it is not present, they are able to use organic compounds for energy.

The *Rhodopseudomonas* genus is one of the commonly found PNSB within EM. These microbes are highly adaptable; they can "grow with or without oxygen, they can use light, inorganic compounds or organic compounds for energy, they can acquire

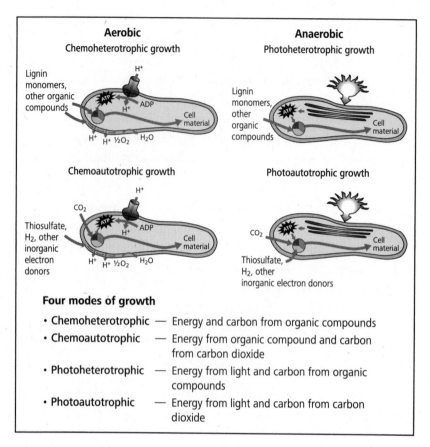

Aerobic

Chemoheterotrophic growth

Lignin monomers, other organic compounds

Cell material

Anaerobic

Photoheterotrophic growth

Lignin monomers, other organic compounds

Cell material

Chemoautotrophic growth

Thiosulfate, H_2, other inorganic electron donors

Cell material

Photoautotrophic growth

Thiosulfate, H_2, other inorganic electron donors

Cell material

Four modes of growth

- **Chemoheterotrophic** — Energy and carbon from organic compounds
- **Chemoautotrophic** — Energy from organic compound and carbon from carbon dioxide
- **Photoheterotrophic** — Energy from light and carbon from organic compounds
- **Photoautotrophic** — Energy from light and carbon from carbon dioxide

Fig. 3.4: Rhodopseudomonas *is highly adaptable and can fuel its growth with four types of metabolic activity.* SOURCE: LE-HOANG, STUDY OF *RHODOPSEUDOMONAS*.

carbon from either carbon dioxide fixation or green plant-derived compounds, and they can also fix nitrogen".[13]

This adaptability allows PNSB microbes to live in a variety of natural locations — they have been found deep in soils, within the digestive tracts of earthworms, and at the bottom of ponds and lakes. PNSB have been used around the world in a variety of industrial applications for bioremediation and the management of chemical and agricultural wastes. They are well-known for their ability to reduce odors by breaking down hydrogen sulfide gas and ammonia. They can also break down the compounds in lignins, the second most abundant polymer on Earth behind cellulose, leading to their use in organic matter recycling. Vinny Pinto says that it appears that PNSB "vastly prefer animal waste products, other wastes or even toxic wastes or pollutants (although they can apparently survive on the wastes and other products of yeast and lactic acid bacteria, and even on the yeast themselves — indeed, that is part of their mutual interdependence and synergy)."[14]

PNSB, yeasts, and lactic acid bacteria are the three main groups of microbes in the consortium that is EM. It is this consortium that gives EM its beneficial effects such as the ability to quickly decompose organic matter, suppress pathogens, and eliminate putrid odors. Now we will start looking into how we can use the EM microbes for our ultimate benefit, mainly fermenting food waste.

CHAPTER 4

How to Make Bokashi Bran

W HENEVER WE COMPOST FOOD WASTE USING BOKASHI, we are essentially fermenting it. In order to control the fermentation process and ensure that it takes place how we want it to, we need to introduce a specific set of microbes to the food waste. In bokashi composting, this is usually done with a product called bokashi bran. Bokashi bran is a carbon source, commonly wheat or rice bran, that has been inoculated with a specific set of beneficial microbes, usually EM. It is usually dried to make it easier to store and transport, but it can be used and stored wet.

In order to make your own bokashi bran, you will need a microbial inoculant. The most common, preferred inoculant is a commercial EM1-type concentrate, or mother culture. This stock solution is sold by a variety of manufacturers with each branding it under their own name. As previously discussed, any EM1-type mother culture should give you similar results as long as it is produced by a reputable manufacturer. In the US, the two main retailers of EM mother culture are SCD Probiotics and TeraGanix and both produce a high-quality, reasonably priced product. I use both products to manufacture my own high-quality EM bokashi bran.

The substrate is typically a carbon source. To make things economical, people typically use industrial food processing by-products such as rice bran, rice hulls, or wheat bran. Can you use other substrates such as paper or coffee? Yes, but the results are not what I would call optimum; more on that later. For now I will focus on making the "industry standard" EM bokashi bran. To do so you need a few basic ingredients, most of which can be found locally (wheat bran and blackstrap molasses), and EM mother culture, which can easily be found online.

How to make EM bokashi bran

Ingredients:
12 lb wheat bran
3 oz EM mother culture
3 oz blackstrap molasses
128 oz water
2 five-gallon buckets

Fig. 4.1: *Water, blackstrap molasses, EM mother culture.*

1. Add 96 oz of water to one of the five-gallon buckets.
2. Boil 32 oz of water.
3. Add 3 oz of blackstrap molasses into the boiling water, stir and pour solution into the five-gallon bucket containing the 96 oz of water.

Fig. 4.2:
Blackstrap molasses, water mixture.

4. Add 3 oz of EM mother culture to the molasses and water solution and stir.
5. Add 12 lb of wheat bran to the other five-gallon bucket.

Fig. 4.3: *Wheat Bran.*

6. Slowly pour the EM mother culture, molasses, and water mixture into the bucket containing the wheat bran.

7. Mix the contents thoroughly with your hand, trying to ensure that all of the wheat bran gets sufficiently moistened.

Fig. 4.4: *All ingredients mixed together.*

8. Overall the mixture should be roughly 50 percent moisture. You should be able to squeeze a ball of it in your hand and lightly toss it up and down without having it break. If you can't get the mixture to form a ball, start adding small amounts of water one cup at a time, mixing, and then trying to form a ball after each cup. If the mixture is too soupy — meaning you can take the ball and squeeze it and have a lot of liquid run out between your fingers, or you can't even form a ball — then slowly add small amounts of dry bran one cup at a time, mixing, and then repeating the test until you get the optimum consistency. It may take a few rounds to get it right, but after you make a few batches, you will get a feel for the optimum consistency.

Fig. 4.5: *Ideal moisture content. The ball of bran is holding its shape.*

9. After all of the ingredients are mixed thoroughly and the moisture content is optimal, add the mixture to a doubled-up plastic bag (two of the same-sized plastic bags, one inside the other). Standard trash bags are convenient, readily available, and work just fine. When all of the mixture has been added, squeeze all the air out of the inner bag and tie it closed. Repeat the process with the outer bag.

   Large ziplock bags or vacuum seal bags also work well. Note: Once this initial mixing is complete, never mix the contents again after they are bagged up, to ensure that the fermenting bran stays anaerobic and oxygen isn't introduced into the mixture.

10. Set the bags aside and let them ferment for at least two weeks in a warm space. Any place that is between 65–90 °F is optimum. The bags should be out of any direct sunlight. Exposure to indirect light is acceptable; the bags do not have to be kept in the dark. It is imperative that you do not open or mix the contents of the bags at all during the two-week fermentation process; the introduction of oxygen at this point is harmful

and will disrupt the anaerobic fermentation process. While the bran is fermenting, the bags may expand because of the gasses created during the fermentation process. If they do expand, no extra action is needed; just let the process run its course.

11. Label, label, label. It is important to label each batch of bran with the recipe you used, including the ingredients and the specific quantities of each. You will also want to know the date that you started fermenting the bran and the approximate average daily air temperature of the area where the bran is fermenting. Tape this label to the bag of bran. After a few weeks, it is really easy to forget exactly what went into a specific batch and when it started fermenting, especially if you are doing longer ferments or using more exotic ingredients. The inputs, air temperature, and fermentation time are important factors to use in the trouble-shooting process. If something goes wrong, then you need to know what you did so you can adjust accordingly. Conversely, if everything goes well, you are going to want to be able to repeat that process again and again. So as insignificant as this step sounds, take a few minutes to right everything down, and label accordingly. It will save you a lot of frustration down the line.

12. After at least two weeks, you can open the bag to check the contents. How do you know if the fermented bran is any good? By smell and by sight. If the smell and visual test check out, then you have usable bokashi bran that can be used to ferment food waste.

Smell Test: The bran should have a sweet and sour yeasty/cheesy/vinegary smell. You may detect a faint alcohol smell. The smell often reminds me of some of the wineries I have been in. Other people have compared it to a sourdough bread dough, which makes sense because the ingredients are similar — yeast,

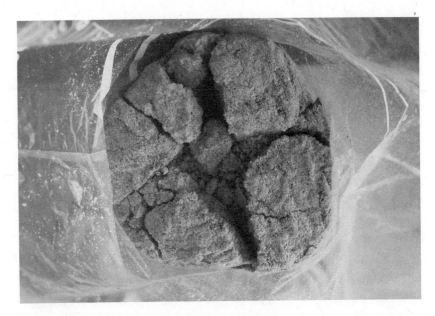

Fig. 4.6: *Finished bran.*

Fig. 4.7: *Finished bran close up.*

sugar, and wheat. The finished bran will smell similar to the ingredients that went into making it. If you notice a musty or moldy odor (think damp basement) or a putrid odor, then the fermentation has failed. The difference in smell between a good and a bad batch of bran is wide, so you should know pretty quickly what you have.

Sight Test: You may see some white mold-like mycelium that has formed on the bran. That is acceptable. If you see any green or black mold, something went wrong during the fermentation process.

If you get a bad batch of bran, a few things could have happened. Most likely air entered the system and the contents didn't have a chance to ferment anaerobically. Other potential causes for the bad mold are the moisture level (the mixture could have been too wet) or poor-quality EM mother culture. *Do not use a*

Fig. 4.8: *Bran ready for drying.*

bad batch of bran in your bokashi system to ferment food waste. You can mix the bad bran into an active aerobic compost pile or bury it in a trench, spray it with a 1:100 diluted AEM solution, and cover it with six inches of soil.

13. Once you have successfully fermented the bran into a usable product, you have two options: you can dry it or you can use it while it's still moist.

    a. To dry the bran, spread it out in an even layer on a tarp or in a container (a kiddie pool works well) and let it air dry. Turn it daily to ensure even drying. The bran should take a few days to dry out. Once dried, it should keep for a year as long as it is stored in dry conditions and out of direct sunlight or extreme heat. I would suggest storing the bran in an airtight container such as a bucket with a gamma lid to keep bugs and moisture out.

    b. You can use the bran while it is still moist. If you do, I suggest doing so relatively quickly unless you store it under controlled conditions. Moist bran can be prone to go moldy (the harmful kind) unless it is sealed perfectly airtight at all times to maintain an anaerobic environment. A bucket with a Gamma Seal works well. I would even take an extra step and keep some plastic pressed down against the surface of the moist bran. When you need more bokashi bran, take the amount you need off of the top layer and avoid digging around through the contents and introducing air into the fermenting bran. As long as the moist bran is kept under anaerobic conditions, its shelf life will be very long, greater than one year. As it sits it will continue to ferment and increase its microbial count and potency over time.

14. Apply the wet or dry finished bokashi bran as needed to your bokashi composting system.

The 12-pound batch process described above is a convenient size for me because the finished bran can easily fit into a five-gallon bucket and will last me a few months. But the process can easily be scaled up or down using the ratios and examples in the chart below.

	Quantities						
Bran (lbs)	1	3	6	12	25	37	50
EM (oz)	0.25	0.75	1.5	3	6	9	12
Molasses (oz)	0.25	0.75	1.5	3	6	9	12
Water (gal)	10 oz.	32 oz.	0.5	1	2	3	4

Unit Conversions

128 oz = 1 gallon = 16 cups = 3.8 l

8 oz = 1 cup = 236 ml

1 oz = 29.5 ml

1 Tbsp = 3 tsp = 0.5 oz

1 lb = 0.45 kg

Can activated EM (AEM) be substituted for EM mother culture?

Yes — if you make a high-quality AEM, then you can use that to inoculate your bokashi bran. This allows you to further extend your EM mother culture, making it even more economical. I have fermented side-by-side batches of bokashi bran using AEM and EM mother culture, and both appeared to turn out the exact same. If there is a difference between the two, then it is beyond what I can detect.

By activating the EM1, you are taking the stock culture and culturing it out one generation. You provide food and the proper conditions for the microbes to multiply. This allows you to take a little EM1 and turn it into a much greater quantity of activated EM, which you can then use.

Activating EM mother culture is a fairly simple and straight-forward process. Each manufacturer and retailer will provide you with its own set of instructions, but the processes are all pretty much the same. A lot of manufacturers recommend a 1:1:20 ratio of EM1 to blackstrap molasses to water. I typically use a 1.5:1:20 ratio, because I think the extra EM1 makes for a stronger culture that gets down to a lower pH faster. EM1 is pretty cheap, so I don't see cost as a prohibitive issue for using more EM1 than less. To make 1 gallon of AEM, my ratio works out to 9 ounces of EM1, 6 ounces of blackstrap molasses, and 128 ounces of water. If I want this to fit into a one-gallon bucket, I just cut back on the amount of water by 15 ounces. This increases the more favorable side of the ratio (EM1 and blackstrap molasses), so in my opinion there isn't any harm in it.

Most people won't want to make a whole gallon of AEM at one time, so a more convenient size is one quart, using the ratio 2.25 ounces EM1 and 1.5 ounces of molasses to 28 ounces of water.

The recipe to activate EM1 using those ingredients is as follows:

1. Heat all 28 ounces of water up to 160 °F (approximately).
   - I suggest using dechlorinated tap water, filtered water, or rainwater. You can use tap water if you let it sit out for 24 hours so the chlorine will dissipate, but in a pinch, regular tap water should do just fine.
2. Add the 1.5 ounces of blackstrap molasses to the water and stir.
3. Let the mixture cool down to 105 °F.
   - At temperatures above 110 °F, you risk killing off some of the microbes in the EM culture, so it is very important to

let the water cool prior to adding EM1 to the water and molasses mixture. The heat was only needed to help the molasses dissolve into the water.

4. Add the 2.25 ounces of EM1 to the mixture and stir.
5. Pour the mixture into a plastic jug and cap.
   - DO NOT USE A GLASS JUG OR A CONTAINER WITH AN AIRTIGHT LID. Gas will be created as the EM is activated. This will create a lot of pressure within the container. A glass or airtight container may burst from the extreme pressure that can build up inside the jug. This can cause injury and make a mess. So use a plastic container with a loose-fitting lid or an airlock.
6. Keep the container in an area that is at least 70 °F at all times. Warmer is better (but no warmer than 105 °F).
   - The container should not be in direct sunlight, but it doesn't have to be in complete darkness either.
   - Warmer temperatures will increase the microbial activity, thereby speeding up the fermentation and pH drop.
7. Burp the jug every few days by loosening the lid, allowing any gas buildup to escape and lowering the pressure inside the jug.
8. When the mixture begins bubbling or after five days, start checking the pH. Continue the fermentation until the pH of the mixture drops below 3.7; then the AEM is ready to use. Typically it will take 7–14 days to get to this stage. At warmer temperatures, this pH drop can occur even sooner.
   - At this stage the AME should have a sweet alcohol smell to it. It should smell more or less the same as EM1 itself.
   - It will probably be bubbling, but not all batches will, especially if they are fermented at the colder end of the range.
   - If a white layer forms on top of the solution, that is fine. It is the yeast.
9. The AEM should be stored at room temperature out of direct sunlight and used within 30 days.

AEM tips:

- Use a container that is roughly the same volume as the amount of liquid you are using. Ferment a quart of the mixture in a quart container, not a gallon container. This eliminates a large air space inside the container.
- You can add different ingredients to the AEM to try to increase its vigor, such as rock dust or sea salt. Add them in Step 4, about 1 tsp of each for a quart batch of AEM.
- Different strategies can be used to insulate the fermentation container and keep the temperature up. Things like aquarium heaters or yogurt makers can be used to raise the temperature of the fermenting liquid. Vinny Pinto goes into a lot of detail on this in his book *Fermentation with Syntropic Antioxidative Microbes*.
- AEM and EM have a variety of other uses besides bokashi composting, such as livestock feed and water additives, odor suppression, and foliar and soil sprays. These topics are beyond the scope of this book, but there is a lot of information out there about the different uses for EM and AEM. I would suggest the information SCD Probiotics and TeraGanix have on their website; Vinny Pinto has also done some great research that is worth reading.

Can EM1 be cultured out more than one generation of AEM?

Short answer: No. As the microbes are cultured out they start to multiply and grow. Over time their ratios will shift away from the original ratio in the EM1. This will result in a culture that doesn't have the same microbial makeup or properties as EM1 or AEM, giving you a mixture of unknowns. When you are trying to achieve a desired end product and specific results, you have no control if your culture is a random, unbalanced mixture of microbes.

Should I use wheat bran or another bran?

Wheat bran is an industrial by-product from the flour-making industry, so it is commonly found in most places around the US (and the world) and is sold in convenient quantities. Smaller quantities can be found in most grocery store aisles and bulk food sections, and large quantities can be purchased at feed stores, where wheat bran is typically sold in 50-pound bags. If you are just starting out, I would suggest starting with a small quantity of bran. After you have perfected the process and hopefully want to make more, then you can scale up.

And remember you don't have to use wheat bran; other carbon sources work as a substrate. Rice bran is another suitable substrate, but it is often more expensive and harder to come by for most people in the US. Many Asian countries utilize rice bran in their bokashi because it is locally and cheaply available as a surplus by-product. I have heard of people using sawdust and wood shavings in their bokashi bran, but I would shy away from that — you don't want to introduce something that high in carbon into your garden soil on a regular basis. Two sources of readily available substrate that you can use are coffee grounds and newspaper. Neither are my preferred ingredient of choice for a variety of reasons, but both seem to work OK.

I occasionally make bokashi bran out of spent coffee grounds, more for experimental than practical purposes. Wheat bran is just easier to work with and is already cheap enough. Also, coffee grounds are a bit tricky to work with if they are wet because you have to estimate their moisture content. Ideally I am looking for around 50 percent moisture, but most of the coffee grounds I collect are wetter than that, so I have to add wheat bran or rice bran to soak up the excess liquid. Once the moisture content is optimal, I then add EM1 and blackstrap molasses, in the same ratios I use for wheat bran. I mix those in with the coffee grounds and add water if needed, then seal the bag up and let it ferment for at least two weeks.

These batches seem to turn out all right. They don't have the same sweet, sour, fermented smell that wheat bran has; the odor is dominated by the coffee. In terms of fermenting performance, the fermented coffee grounds handle the kitchen waste properly and the final bokashi pre-compost turns out fine. So coffee grounds appear to be a viable alternative, and if you have an unlimited supply of them, it would probably be worth perfecting the process. Another option is to cut your wheat bran with a smaller amount of coffee grounds. That way you extend your wheat bran a little bit and maintain the predictability that comes with it. In terms of the coffee grounds themselves, I like to use the straight grounds, no filters. It isn't that the paper filters will cause problems, but they are just a pain to work with in the manufacturing and final product stages, so I keep them out for the fermentation.

Newspaper bokashi bran

I have made and used bokashi bran from newspaper. It ferments the food waste well, but I think handling it is a pain. The sheets of newspaper tend to stick together and clump up in the bucket, which requires a lot of ripping, cutting, or shredding to get them into small pieces to better mix with the food waste. In my opinion wheat bran produces a better final product that is easier to work with while still being cost effective. But people like to save money and make things, so here is how I make newspaper bokashi bran.

After much experimenting, I have found that this recipe works for me. The ratio of molasses and EM are high, at 1:1:3 (EM, blackstrap molasses, water), so the savings you make by using paper over wheat bran are offset by using more EM.

½ cup EM
½ cup blackstrap molasses
1½ cups water
A stack of newspaper (not the glossy ads)

1. Mix the three ingredients together using the same methodology as with wheat bran.
2. Pour the mixture into a sheet pan or similar container.
3. Soak a stack of newspaper in the mixture for at least an hour.
   - How much paper is hard to say; I would start with about ¼" or so. You can always add more later.
   - Flip and shuffle the stack of paper periodically to ensure a more uniform absorption.
4. After an hour, remove the soaking newspaper, shake off the excess liquid, and place it inside a large ziplock bag. Remove the extra air from the bag and seal it.
5. Keep the sealed bag at room temperature and out of direct sunlight for at least two weeks to allow the contents to ferment.
6. After two weeks, you have a couple of choices; you can use the newspaper wet in your bokashi bucket or you can dry it.
   a. If you use the newspaper wet, be sure to reseal the bag and remove excess air each time you open it. I am not sure of the lifespan of this newspaper bran so I would suggest using it within 30 days.
   b. If you choose to dry the newspaper, separate the sheets as best as possible. Allow them to dry. Then shred or cut them up if desired and store them in dry conditions out of direct sunlight. I am not sure of the lifespan of this newspaper bran so I would suggest using it within three months.
7. Use the newspaper bran in your bokashi bucket similar to how you would use wheat bran bokashi bran.
   - Add a layer of newspaper bokashi bran between each layer of organic waste. The layers of organic waste should be no more than one inch thick.
   - Compact each layer after you add the waste and the newspaper bran. This removes air and ensures more surface contact.
   - Allow the contents to ferment for at least two weeks, as with wheat-based bokashi bran.

Molasses, the sugar source

In order to get the microbes to multiply and proliferate, we need to give them something easy to eat — molasses, more specifically blackstrap molasses. Blackstrap molasses is the final concentrated by-product from the third boiling of the sugar refining process. It is a carbohydrate-rich "waste" product that contains a large amount of vitamins, minerals, and trace minerals that aren't found in white refined sugar. The extra nutrients in the blackstrap minerals give the microbes a more complete food source, resulting in a stronger consortium than you would get from simpler sugars. Vinny Pinto has done a lot of work experimenting with various sugar sources and has found that simpler sugars result in problems during fermentation (such as the pH dropping too rapidly) and a short AEM shelf life. Blackstrap molasses is also the sugar source recommended by the two main EM manufactures in the US, TeraGanix and SCD Probiotics.

Blackstrap molasses can be purchased from a variety of sources. Small quantities can be found in grocery stores, and large quantities (usually in five-gallon sizes) can be purchased from local feed stores. Buying bulk is usually a lot cheaper than buying small quantities at a grocery store, so if you think you will do any sort of longer-term brewing of AEM or making of bokashi bran, I would suggest going the five-gallon route. The larger quantity doesn't cost that much more, and there are a lot of garden uses for molasses (soil drench, compost teas, etc.) so the extra won't ever go to waste.

Should I use organic blackstrap molasses?

No, but you want to be sure that you are buying pure molasses. You want to avoid any molasses that has been sulfured or has had any other preservatives or additives added to it. Anything that is added to the molasses is going to interfere with the multiplication of the microbes and with the fermentation of the bran. So always be sure that you are buying pure molasses, organic or not, regardless of the source. If you are in doubt, ask.

It is also important to note that molasses may have a pH buffer mixed into it to help preserve it. This can interfere with the fermentation process when you are activating EM or fermenting bokashi bran. So if you do use an organic product, you may experience atypical results.

What type of water should I use?

Tap water is fine. I would suggest letting it sit out for 24 hours to give it a chance to dechlorinate, but you don't have to. Unfortunately letting the water sit out in the open won't allow the chloramine, which is common in a lot of municipal water with or without chlorine, to dissipate. There really isn't an easy way to remove the chloramine without using other chemicals or filters, but again, while not favorable, the presence of chloramine doesn't render the water unusable. The concentration of microbes in the EM mother culture is extremely high, and there aren't enough chemicals in municipal water to make a significant difference in their multiplication rate. But, like with anything else, high-quality water is ultimately more beneficial than low-quality water. If you have a cheap and easy source of spring water, then I would use that. Most people don't, so they turn to filtered water. Filtered water is controversial because of the processes it is put through; some believe that these result in an unstructured or "dead" water. If you don't subscribe to those theories, then don't worry about it and use filtered water. I have fermented a lot of bokashi bran using filtered water, and they all seemed to turn out fine. If you don't want to use filtered water and municipal water is out, then you are looking for a spring. I would also suggest staying away from any sort of pond, stream, or lake water unless you know that it is of high quality. But overall the process if pretty forgiving because of the very high concentration of powerful microbes it uses. This and a readily available food source (blackstrap molasses) ensure very fast replication and a robust culture.

Should you dry the bran in direct sunlight?

There are mixed thoughts on this, and I am not convinced that there is a correct answer. I typically dry my bran out of direct sunlight in my garage. Sun-dried bran appears to work just as well, so it is really hard to detect any measurable negative effects from the sunlight. The jury is still out on this one.

Why buy EM mother culture when I can just make my own inoculant?

Some people will complain that EM1 mother culture is expensive, and I will beg to differ. Current market prices are roughly $25 for 32 ounces. If you factor in tax and shipping then you are up to a dollar an ounce. Those 32 ounces can then be extended to make activated EM (AEM) at a ratio of 20 to 1, resulting in 640 ounces of AEM. If you just stopped there you would have five gallons of usable product at a cost of six dollars per gallon — very reasonable for an agricultural product.

Then consider that the AEM is further diluted down for various uses at ratios such as 40 to 1; the five gallons turns into 200 gallons, bringing the cost down to 16 cents a gallon. Needless to say that is very cheap, and the cost-prohibitive argument of EM1 becomes a moot point. For a relatively small amount of money, you can brew a very large amount of an incredibly powerful microbial inoculant that you can use to make your own bokashi bran or for a variety of other uses.

Still, some people will feel the need to venture down the road of making their own inoculant. While I applaud the effort, I think that energy can be better spent in different ways. EM1 is a pretty sophisticated collection of microbes, and it took some complex processes to get them all to function together in consortium. This isn't something that can easily be recreated in a home situation even under the best conditions, so let's be clear — no one will be making their own EM1 consortium. What most people will end up making is a lactic acid bacteria (LAB) solution, which again isn't EM1, it is a lactic acid bacteria solution; there aren't any yeasts or PNSB, and there aren't any of the other supporting members of the consortium; it is a single-microbe solution. Yet people will still call this solution homemade EM. It is the most common homemade solution you see going into homemade bokashi bran recipes.

OK, so will it work? Yes, but the results will be slightly different than with EM bokashi. The lactic acid bacteria will be able to ferment the food waste, but you will be losing some of the digesting and odor-reducing power that the other microbes in EM bring to the table. I have even read articles where people discuss using plain, untreated wheat bran to ferment organic waste, with positive results. If you do use LAB bokashi, err on the side of using more rather than less. The starter culture is homemade, and there is no way to know how powerful it was to begin with (compared to EM, which is a known), so a little extra helps your chances of success.

How to culture a homemade *Lactobacillus* (LAB) solution

1. Mix one part rice thoroughly with two parts water. Stir the mixture vigorously.
   - Any type of rice will work.
   - I suggest using a dechlorinated tap water, filtered water, or rainwater. You can use tap water if you let it sit out for 24 hours, allowing the chlorine to dissipate, but in a pinch, regular tap water should do just fine.
2. Pour the rice wash water through using a strainer or cheese-cloth to separate off the solids. Keep the liquid rice wash

water, which should be cloudy in color, and discard and compost the leftover rice and solids.

Fig. 4.9: *Two parts water, one part rice.*

Fig. 4.10: *Rice wash water.*

Fig. 4.11:
*Straining the rice
wash water.*

3. Pour the rice wash water into a container with 50 percent head space, allowing plenty of air to circulate.
   • I suggest using a glass or plastic container. Metal containers may cause issues.
4. Cover the container with a paper or cloth towel. The cover will keep insects out and should allow air to move in and out.

Fig. 4.12:
*Rice wash
water, covered
with 50 percent
head space.*

It must not be airtight because we want bacteria to be able to enter the container, which they can do through a towel.

5. Place the container out of direct sunlight at room temperature for seven days.

6. After seven days the mixture should smell sour. Filter the liquid through a strainer or cheesecloth. Keep the resultant liquid. Discard and compost any solids that were filtered out.

   • During the seven days, a whole collection of "wild" microbes will have grown in the liquid using the rice wash as a food source. We are ultimately after just the *Lactobacillus*, so we have to isolate that within the mixture.

7. Add ten parts milk to one part of the strained rice wash solution, allowing plenty of head space in the container so air can circulate.

   • Any type of milk will work, but I would tend to stay away from the ultra-pasteurized milks because they could cause issues with the process.

Fig. 4.13: *10 parts milk, 1 part strained rice wash solution.*

Fig. 4.14: *Milk, strained rice wash solution combined. Ready to sit for 7 days.*

8. Cover the container with a paper or cloth towel. This will keep insects out and should allow air to move in and out of the container. It should not be airtight.
9. Place the container out of direct sunlight at room temperature for another seven days.
10. During these seven days, most of the solids in the milk (fat, proteins, carbohydrates) will float to the top of the container,

Fig. 4.15: *After 7 days, the solids and serum have formed.*

resulting in a "cheese." That can ultimately be discarded and composted. Beneath the cheese, you will see a yellowish serum; that is our lactic acid bacteria solution — what we ultimately want.

- Remove the solids with a slotted spoon and discard. If necessary you can remove any remaining solids by filtering through a strainer or cheesecloth. The remaining yellowish-colored liquid is a *Lactobacillus* serum.

Fig. 4.16: *A close up of the solids, "cheese."*

Fig. 4.17: *The LAB serum with the solids removed.*

Fig. 4.18: *A close up of the LAB serum.*

The resultant *Lactobacillus* serum can be stored in two different ways — with or without refrigeration.

a. You can store the *Lactobacillus* serum in the refrigerator in its pure form. Just pour it into a container and place it in the fridge. The container doesn't have to be airtight. If stored this way, the pure LAB serum will stay viable for well over one year. I would suggest using it sooner, but it will last if needed.

b. To store the LAB serum at room temperature without refrigeration, just mix it with equal parts molasses. Keep this mixture out of direct sunlight and out of extreme temperatures. Keep the container sealed to keep insects out (it doesn't have to be airtight). If kept in these conditions, then the mixture will stay viable for well over one year, but sooner use is suggested.

Fig. 4.19: *Stored as a pure culture, in the refrigerator.*

Fig. 4.20: *Mixed with blackstrap molasses to be stored outside of the refrigerator.*

The homemade pure LAB serum can be used to make bokashi bran using a ratio of 2:1:10 (LAB serum, blackstrap molasses, water). Then just follow the same process as above for EM bokashi bran.

The homemade LAB solution has a variety of uses. Gil Carandang has done some interesting work on this topic, which he describes in his book *Grow Your Own Beneficial Indigenous Microorganisms and Bionutrients in Natural Farming*. It is worth reading. Another great resource is the book *Natural Farming* by Cho Han-Kyu and Cho Ju-Young.

CHAPTER 5

The Fermentation Vessel and How to Make Your Own

WHEN COMPOSTING WITH BOKASHI, the key ingredient besides the microbes is the fermentation vessel (bokashi bucket). Since we are fermenting food scraps anaerobically, it is critically important that the vessel we use can be sealed hermetically and can maintain that seal over a period of time. There are a variety of different containers that can be used, large and small, homemade and commercial; all will work fine as long as they stay airtight.

Commercially available bokashi buckets are all basically the same; they just have slight design differences to try to improve ease of use, functionality, and longevity. All have a lid that can seal hermetically, a riser, and a screen above the bottom to allow the bokashi tea to drain through while keeping the solid contents on top; most also have a spigot which allows the user to easily drain the bokashi tea from the bucket. There is nothing hi-tech going on here. The buckets are designed for ease of use and convenience, ensuring that you actually use them and end up with a good finished product.

Here is a picture of a standard commercial bokashi bucket that is popular in the US.

Fig. 5.1:
*Commercial
bokashi bucket.*

This bucket has a flexible snap-style lid which can be removed easily, a plastic screen in the bottom of the bucket which divides the fermenting contents from a recess at the bottom used to collect the bokashi tea, and a spigot which can be used to access the tea. I have used it for over three years and have never had any issues with it. It still seals fine and doesn't leak. The plastic is pretty durable, and as long as it is kept inside and out of the sun, it should last for a long time.

This leads us into the plastic discussion. What if you don't like plastic? Then you probably won't be fermenting food waste with bokashi, because when it comes to commercially available bo-kashi buckets, there aren't any non-plastic alternatives out there. So if you want to go plastic free, you will have to make your own

Fig. 5.1-1: *Bokashi bucket lid and divider screen.*

Fig. 5.1-2: *Flexible snap-style lid.*

or adapt some existing technology. Cutting the top of a glass car-boy might be a start. Ceramic vessels with water trough airlocks are another option; these have been used for fermentation for thousands of years. You won't get the potential toxins that come with plastic, but your tradeoff will be durability and weight. I have also heard of people using glass storage jars with an airtight bail and seal to ferment their food waste. That is too small for me, but it is an alternative to plastic.

While I completely agree that getting as much plastic as pos-sible out of our daily lives is a good thing, bokashi composting is an area where plastic is allowing something eco-possible to hap-pen on a large scale, so the overall benefits surely outweigh the negatives. Most of the current commercially available bokashi buckets are food-grade plastic and use recycled plastic in their construction. So while plastic isn't the best option in the realm of all things, it is currently the only option. I strongly encour-age anyone out there to create a commercially distributable, yet profitable and affordable, non-plastic bokashi bucket. It would be awesome, and you would probably sell a lot, because I would buy it and a lot of others would as well.

Returning to the world of plastic, it is possible to make your own usable, effective bokashi bucket using common hardware store ingredients. There is one main thing you need to think about before you decide on a design: do you want to collect and use the bokashi tea? If yes, then you will need to use a bucket design that allows you to easily do that. If not, then no worries, even easier. If you are unsure, there is even a design to cover the in-betweeners.

The double-nested bucket

If you are not sure that you want to collect the bokashi tea over the long run, but you might want to sometimes, then here is the bucket design for you: the double-nested bucket.

There isn't any hi-tech engineering going on here. It is just what it sounds like, a bucket inside of another bucket. This is the absolute easiest setup for anyone who may want to collect the bokashi tea. All you need is an airtight lid and two five-gallon buckets. Take one of the buckets and drill a bunch of holes in the bottom. Hole size and quantity doesn't matter; you just want them to be big enough so they don't clog easily and allow for sufficient drainage; ¼" is probably your best bet. Make sure the holes are evenly distributed around the bottom of the bucket. Again, there isn't any serious engineering going on here, we are just drilling holes in a bucket. You can always add more holes or drill them out bigger if needed, but you can't go the other way, so err on the side of fewer smaller holes.

After all of the holes are drilled, all you have to do is put the bucket with holes inside the other five-gallon bucket with no holes in it. The two buckets will nest together perfectly, leaving an air gap at the bottom between the two buckets where the

Fig. 5.1-3: *Holes drilled into the bottom of the inside bucket.*

bokashi tea can collect. When you want to harvest the bokashi tea, all you need to do is to remove the upper bucket and pour out the tea from the lower bucket.

Fig. 5.1-4: *Bokashi tea collected in the bottom bucket.*

This system removes the hassle of dealing with spigots, which can be leaky, difficult to install, and moderately hard to locate and purchase for some people. Also if you ever decide to stop collecting the bokashi juice, just remove the bucket with holes in it and you can ferment your food waste in the remaining intact bucket.

Is this the easiest way to harvest the juice? No, a spigot is probably easier, but this type of setup is super simple and most people have a few five-gallon buckets sitting around, so the only item you may need to buy is a lid for the upper bucket.

If you are going to make your own bokashi bucket then I would strongly suggest using a Gamma Seal for the lid.

The Gamma Seal will give you an easy-to-use, long-lasting, high-quality lid that is designed to be opened and closed thousands of times using only one hand. And the lids are designed to

Fig. 5.2-1:
*Gamma Seal
threaded bucket
lid.*

Fig. 5.2-2:
*Gamma Seal lid
on a bucket.*

be airtight, with the two pieces of the lid threading together and sealing against a rubber gasket. The Gamma Seals are relatively cheap at around ten dollars each and are sold at many hardware stores or online. They are heavy duty, so they will last for an extremely long time, and given their relative ease of use, they will save you a bunch of frustration in the long run for only a few dollars more of initial investment.

If you don't want to use a Gamma Seal for whatever reason, then you have a variety of alternative lids. There are the cheap press-on style lids, which are sold at big box hardware stores for about two dollars each. These are not always airtight and will tend to crack and become deformed after a while, thereby losing their ability to seal the bucket airtight. There are the tab-style lids, which are often found on food-grade buckets. These lids are airtight initially and do incorporate a rubber seal, but in general they are extremely difficult to take on and off if you seal them properly each time. If you have to remove the tab-style lids on

Fig. 5.2-3: *Cheaper lid alternatives: press-on (left), tab-style (right).*

a daily basis, then you will understand why a Gamma Seal is so great. So spend the extra few dollars and get a Gamma Seal. I feel pretty confident in saying that once you go down the Gamma Seal route, you will never go back.

The spigot bucket

Another option for collecting the bokashi tea using a homemade bokashi bucket is to install a spigot on the five-gallon bucket. Spigots similar to the one below are available at homebrew supply stores and online.

They are relatively cheap and make harvesting the bokashi tea a little easier. They come in especially handy if you have multiple bokashi buckets going, because you can stack the buckets on top of each other and still be able to access the tea in each bucket without having to take the whole stack apart.

The spigots themselves are pretty simple to install. The first thing you will need to do is choose the spot on the bucket where

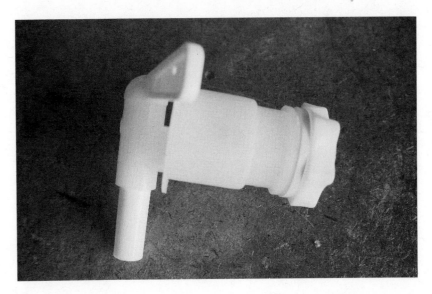

Fig. 5.3: *Plastic spigot, 1".*

you want to install the spigot. It sounds simple — just put it at the bottom. But you do have to take one thing into consideration, the height of the spigot spout and how far the spigot is installed from the bottom of the bucket. If you install the spigot too close to the bottom the spout on the spigot is going to end up lower than the bottom, of the bucket. This means that you can never set the bucket on a flat surface; you are always going to have to hang it over the edge of something so the spigot doesn't rest on the ground or you will have to cut the spout on the spigot to accommodate.

When you install the spigot, take this into account. Just hold the spigot up to the side of the bucket before you drill the hole so you can see if the placement will be acceptable to you. Too high is going to be better than too low; you can always tilt the bucket to get the bokashi tea out, but you can't move the hole up if it is drilled too low. Once you have aligned the spigot, you just

Fig. 5.4: *Spigot installed too low. The bucket cannot sit flat.*

Fig. 5.5: *Spigot installed properly. The bucket can sit flat. Note the improperly placed hole.*

have to drill the hole using a spade bit or hole saw and install the spigot. Most spigots have a standard diameter on their neck, such as one inch, which makes it easy to drill out with a correspondingly sized drill bit. Be careful when you are drilling the hole because five-gallon buckets are prone to cracking around the drill hole, and if that happens, you will have to start over. Make sure to install the rubber gasket on the *inside* of the bucket; this will prevent any leaks (assuming that you sized your hole properly). Then tighten the nut on the spigot, put the lid on the bucket, and you are good to go — you can start fermenting food waste in your new bokashi bucket.

What about the air gap created by the false floor on the double-nested buckets and the commercial available buckets (the spigot bucket doesn't have one)? Where will the bokashi tea collect so I can harvest it easily? In most cases you won't need an air gap, or area at the bottom of the bucket for bokashi tea to collect. The organic matter packed into the bucket will generally be permeable enough to allow bokashi tea to collect and flow out the bottom of the bucket without clogging the spigot. But if you do have problems with the spigot clogging, you can try a few things on your next batch of kitchen waste to avoid that problem. You can create your own air gap and false floor at the bottom of the single bucket by putting something inside the bucket. You can use a commercially available product called a Grit Guard, which will give you a permeable raised floor, allowing the tea to flow through but keeping the large organic material on top. You can go the homemade route and use something such as a plastic microwave splatter guard. This will fit right into the bucket, giving you the separation that you want. Splatter guards are easy to find and are usually available at dollar stores, so cost isn't an issue.

Fig. 5.6: *Microwave splatter guard.*

Fig. 5.7: *Microwave splatter guard used in the bucket as the divider screen.*

If you don't want to collect the bokashi tea, then you can ignore the spigot and just use a single-bucket system. All you need is a five-gallon bucket and an airtight lid. The process will be described in more depth later, but basically you place a dry material such as newspaper or dry leaves at the bottom of the bucket to absorb the excess liquid. The bokashi tea isn't separated out and is just processed with the bokashi pre-compost after the fermentation process.

All of these bokashi fermentation systems talk about using five-gallon buckets. There are a few reasons for that: five-gallon buckets are available everywhere, often free, and their size and filled weight make them practical and easy to handle for most people. Many chain restaurants purchase food products and sauces that are delivered in five-gallon buckets; once empty the restaurants typically throw the buckets out. Ask around or check some dumpsters and you can probably find more food-grade buckets than you can use.

You aren't limited to the five-gallon size, and oftentimes creativity is the only thing limiting your design. Bokashi fermentation systems can easily be scaled up or down to any size container. If you are generating enough waste, you could use a fifty-plus-gallon plastic drum or a rollaway trash can, though if you do size up, keep in mind the final weight of the filled container — a fifty-plus-gallon drum filled to the top with moist organic waste is going to be extremely heavy and hard to move around and empty. From a bucket management standpoint, a number of smaller containers may make more sense than one large container. That being said, there are systems out there which utilize fifty-gallon plastic drums and rollaway trash cans in more industrial settings like restaurants, because of the extreme amounts of kitchen waste that are being generated on a daily basis.

One factor to consider when you are sizing your system is how much waste you generate. I recommend a bucket that takes no more than two weeks to fill (less time is even better). Bokashi fermentation needs to take place anaerobically, and if you keep opening the bucket to add fresh food waste you keep introducing oxygen to the system. The more you do this, the more the final fermentation gets delayed and the higher the chances that the contents will go rancid and spoil because the anaerobic microbes can't dominate. Experience tends to show that when the constant opening and closing drags out for more than two weeks you will probably ending up with a bad batch (unsuccessful fermentation). So if you can't fill the container you are using to at least 80 percent of its maximum capacity within two weeks, the fermentation container is too big and you should use a smaller one. In

addition to reducing the amount of times that oxygen is introduced to the system, this will also decrease the airspace between
the lid and the top of the kitchen waste. Ideally this airspace will
be as small as possible, meaning the bucket is jam-packed right
to the lid.

Regardless of container size, the contents of the bokashi bucket should be packed down extremely tight — as tight as you can
get them. Some retailers offer a potato-style masher to do this. I
have found the easiest way to pack down the contents is to use an
old plastic grocery bag. After you put in the first round of scraps,
add the bag to the bucket. Just lay it flat on top of the contents;
it will be the perfect size for most retail bokashi buckets and
homemade five-gallon buckets. The grocery bag will then create
an oxygen barrier separating the fermenting food waste from the
air gap above and will also give you a clean surface to compact
the waste with. Nothing fancy here, just lay the bag flat and press
down with your body weight. You can get an amazing amount of
waste into a five-gallon bucket if you compact it this way.

For a typical family, the five-gallon size is optimum; most fam-
ilies will fill one in about a week. Having four separate buckets in
rotation creates a continuous system that constantly generates an
organic soil amendment and is always available to receive new
kitchen waste. Four commercial bokashi buckets aren't that expensive if you want to go that route. If you go homemade, the four
buckets will be really cheap, especially if you can scavenge them
from somewhere. The low-cost homemade option also allows
you to easily scale up to even more buckets if needed and desired.

At this point, if you really want to try bokashi composting
and are willing to do a little work, cost is no longer an issue. For
less than fifty dollars you can build your own bokashi bucket, ferment your own bran, and still end up with a bottle of EM1 that
has a variety of other non-bokashi uses all to itself. Building the
bucket and making your own bran will also help you to better

understand the whole process and get a better feel for things. You might even make some discoveries along the way that can ultimately make things easier for yourself and other bokashi enthusiasts. The space- and wallet-friendly options make bokashi fermentation a practical and viable choice for people who currently won't or can't compost traditionally for whatever reason, giving the world another solution for diverting food waste from the landfill to the soil.

CHAPTER 6

How to Compost Your Kitchen Waste with Bokashi

IT IS IMPORTANT TO REMEMBER THAT BOKASHI COMPOSTING involves working with living systems — specifically, a symbiotic collection of microorganisms used to break down organic matter by fermenting it into a living, beneficial soil amendment while limiting harmful by-products. These microorganisms are doing this in the "wild," outside of a laboratory, so conditions are different in each and every situation and results will always vary to some extent. You can generally expect the same thing to happen every time, but it won't always happen exactly as planned because the variables — temperature, kitchen waste type and mixture, moisture level, bran quality — are always slightly different. It is our job as composters to watch the changes and results and make adjustments accordingly. Overall the process is very forgiving, and it is pretty hard to totally screw it up. And even if you screw up the fermentation the organic matter will still end up feeding the soil.

The first time that you compost your kitchen waste don't expect to open the bokashi bucket and see black or dark brown compost, the image most people conjure up when they think of compost. This process is different, and the end result is different too. Remember that composting with bokashi is a fermentation process, so the material comes out of the bucket looking more or less similar to how it looked when it went in (think pickles or *kimchi*). It only becomes rich black humus when it is mixed with soil and further broken down by the soil biota. When we ferment food waste, we are simply speeding up the front end of the process.

It is important to note that anytime you compost, regardless of the method, input size matters; smaller objects will decompose faster than larger objects. By decreasing the size of something, you increase the total surface area that is exposed to the various decomposers. Also, cutting a material open gives the decomposers a foothold. Fruits and vegetable skins have natural compounds in them to help fight off decay, so cutting them open bypasses the protective skin and gives the microorganisms direct access to something they can start eating and breaking down. The protective ability of vegetable peels is pretty impressive. I have seen whole apples and tomatoes last for a few weeks in very active worm bins and look relatively unchanged. Squeeze the fruit and pierce the peels and they get eaten pretty rapidly. So when dealing with the scraps you will put into your bokashi bin, smaller and cut up is better than larger and whole.

But cutting stuff up takes extra time, why bother?

A lot of bokashi users compost on a fairly small scale, so a little extra time on the front end isn't an issue. Many users live in apartments or houses and are only interested in composting their kitchen waste, so the amount of material isn't enough to deter a little extra processing. Breaking food waste down mechanically before it hits the bokashi bucket will only take a few extra knife strokes and an extra minute or two when you are preparing a meal, but the smaller size really helps to speed up decomposition on the back end. This plays to the desires of smaller users, who tend to want to cycle things faster. Many are also gardening in small spaces and want to go from food waste to usable compost fairly quickly. Preprocessing the waste can help speed up this cycling.

If you don't care about speed, fine. You don't have to take any extra measures to decrease the size of anything. It is all a time trade-off. Less time breaking down food waste on the front end means it will ultimately take longer to ferment and break down on the back end. Larger operations such as farms and institutions might not see the need to spend a lot of time and effort preprocessing waste because they have the space and time to allow for more composting to take place. Large-scale users who want to maximize their fermentation speed and efficiency can use industrial shredding machines that can process large volumes of food waste before it goes into a bokashi composting system or a vermicomposting operation. These are expensive and use grid-based energy, so there is a trade-off. Every individual will ultimately have to decide what they need for their system.

To ferment your food waste, you will need to introduce a microbial source. The microbes are usually applied on a solid carrier, commonly called bokashi bran. Earlier you learned how to make your own bran. Now I will explain how to use that bokashi bran and your bokashi bucket (commercial or homemade) to ferment your kitchen waste. The process is extremely simple, and even a child could do it.

How to use your bokashi bucket with a spigot or drain:

1. Sprinkle a layer of bokashi bran onto the bottom of the bucket.
   - A dusting is all that is needed.

Fig. 6.1: *Start with a layer of bokashi bran.*

2. Add a layer of kitchen waste. The layer should be about one inch thick.
   - Smaller pieces of waste are better.

Fig. 6.2: *Add a layer of kitchen waste.*

3. Sprinkle a handful of bokashi bran on the layer of kitchen waste. A light dusting is all that is needed; you do not need a solid layer of bran.
   - If your bucket contains meat, sprinkle extra bran in that area. It is better to add too much bran than too little.
   - If you start to smell any bad odors when you add scraps to the bucket, then you need to start adding more bran.

Fig. 6.3: *Scraps covered in bokashi bran.*

4. Lightly stir up the contents and compress them.
   - Compressing the contents ensures adequate contact between the food scraps and the microbes on the bokashi bran. It also helps to remove any air spaces between the materials and forces any extra liquid in the system to drain lower.
5. Cover the layer of kitchen waste and bokashi bran with a plastic bag and press down.

- This step isn't mandatory, but I have found that it helps the process to keep scraps covered until the bucket is completely full.
- Another optional step: Place a brick, a bag of stones, or another heavy object on top of the bag to help compress the layers.

Fig. 6.4: *Compress the kitchen waste. An old plastic bag comes in handy.*

6. If you are adding another layer of scraps at this time, repeat steps 2–5: if you are done adding scraps, continue to step 7.
7. Put the lid back on the bucket it and keep it somewhere warm (60–90 °F) and out of direct sunlight until the next time you add scraps.
8. Add more scraps using steps 2–7 until the bokashi bucket is completely full.

9. Once the bucket is full, seal it and place it somewhere warm (60–90 °F) and out of direct sunlight. Let the bucket sit for at least two weeks without opening it.

    • Cooler temperatures may require a longer fermentation time, while warmer temperatures may require as little as one week. Judge your results and adjust accordingly.

10. Label and date the bucket.

    • As insignificant as it sounds, this is an important, often-forgotten step. There have been many times in the past when I have completely forgotten when I sealed a bucket. I have also mixed up similar-looking buckets and got confused about which was which. If you are experimenting with different formulations of bran, labeling and dating is especially important. To achieve any sort of consistency, you need to be able to evaluate your results and troubleshoot accordingly. Writing each formula on a label takes care of everything. No more guesswork or trying to remember; it is what it says it is.

11. Drain the bokashi tea every other day and use it immediately.

Fig. 6.5: *Draining the bokashi tea.*

12. After the two weeks have passed you can process the bokashi pre-compost accordingly. More on that in Chapter 7.

How to use your bokashi bucket without a drain

You can ferment food waste in a bokashi bucket without a drain as long as you take some extra measures to deal with the liquid that will accumulate at the bottom of the bucket. Too much unabsorbed standing liquid can cause problems. This can be mitigated by adding a layer of dry material such as shredded paper to the bottom of the bucket to absorb any extra liquid.

1. Sprinkle a layer of bokashi bran onto the bottom of the bucket.
   - A dusting is all that is needed.
2. Add three inches of shredded or crumpled paper to the bottom of the bucket.
   - This assumes that you are using a five-gallon bucket. If you are using a different size bucket, you just need a substantial amount of dry material. Keep in mind that it will compress and compact as it gets wet and weight is applied.
   - Bread, old rice, or such will also work fine.
3. Sprinkle a layer of bokashi bran onto the dry absorbing material.
   - A dusting is all that is needed.
4. Add a layer of kitchen waste. The layer should be about an inch thick.
   - Smaller pieces of waste are better.
5. Sprinkle a handful of bokashi bran on the layer of kitchen waste. A light dusting is all that is needed; you do not need a solid layer of bran.
   - If your bucket contains meat, sprinkle extra bran in that area. It is better to add too much bran than too little.
   - If you start to smell any bad odors when you add scraps to the bokashi bucket, then you need to start adding more bran.
6. Lightly stir up the contents and compress them.
   - Compressing the contents ensures adequate contact between the food scraps and the microbes on the bokashi

bran. It also helps to remove any air spaces between the materials and forces any extra liquid in the system to drain lower.

7. Cover the layer of kitchen waste and bokashi bran with a plastic bag and press down.
   - This step isn't mandatory, but I have found that it helps the process to keep scraps covered until the bucket is completely full.
   - Another optional step: Place a brick, a bag of stones, or another heavy object on top of the bag to help compress the layers.

8. If you are adding another layer of scraps at this time, repeat steps 4–7; if you are done adding scraps, continue to step 9.

9. Put the lid back on the bucket it and keep it somewhere warm (60–90 °F) and out of direct sunlight until the next time you add scraps.

10. Add more scraps using steps 4–8 until the bokashi bucket is completely full.

11. Once the bucket is full, seal it and place it somewhere warm (60–90 °F) and out of direct sunlight. Let the bucket sit for at least two weeks without opening it.
    - Cooler temperatures may require a longer fermentation time, while warmer temperatures may require as little as one week. Judge your results and adjust accordingly.

12. Drain the bokashi tea every other day and use it immediately.

13. After the two weeks have passed you can process the bokashi pre-compost accordingly. More on that in Chapter 7.

When you open the bucket after the two weeks, you will encounter a sour, cider vinegar-like smell. Contrary to what you might read, the bokashi composting process does smell, though it should smell closer to pleasant than bad. As long as the bucket smells vinegary, not rancid or putrid, then fermentation has succeeded. How will you know if it is smells bad? It will smell bad — there is no mistaking the difference between a successful sour, cider vinegar scent and a putrid spoiled stench.

After a successful ferment, you will most likely see some patches of white cotton-like mycelium covering the surface of the food scraps. White mycelium is good, and a sign that things went well. You will probably see it, but not always; as long as the bucket smells all right, the fermentation was successful. The appearance of white mycelium is good, but not a mandatory sign of a successful fermentation.

Various pictures of finished bokashi pre-compost.

Fig. 6.6.

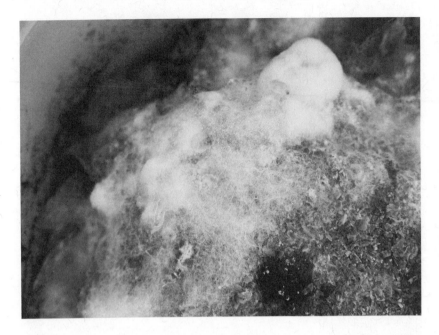

Fig. 6.7.

Fig. 6.8.

Fig. 6.9.

CHAPTER 7

Using the Fermented Food Waste

TWO WEEKS HAVE NOW PASSED, and the microbes inside your bokashi bucket have been busy fermenting all of your food waste, turning it into bokashi pre-compost. Now it is time to finally open the container and put the fermented scraps to use in the soil. Don't expect to find the same dark black, earthy-smelling humus compost you find in a traditional aerobic compost pile. What you will find is fermented (pre-composted) food waste that looks very similar to how it looked when it went into the bucket. The fermentation process doesn't change the outward appearance of the scraps, just as cucumbers don't change appearance when they are pickled. But the physical and chemical structures of the scraps have changed during the fermentation process. The microbes that you inoculated the food waste with have fermented and pre-composted it so it can be rapidly assimilated into the soil.

So what do you do with all this fermented food waste, or bokashi pre-compost?

The most common method for incorporating it into the soil is the trench method. Someone that is building their garden soil or has a patch of land they are trying to rehabilitate would use this method. It makes one big assumption: that you have access to land to dig a trench (and it isn't frozen). If you don't have any digable land, don't worry; landless methods will be covered a little later in this chapter.

The first step is to dig a twelve-inch-deep, shovel-width trench in your garden. It should be twice as long as your container is high, so if you are using a fifteen-inch-high bucket, dig a trench thirty inches long; exact dimensions aren't critical, so there is no need to get out a tape measure.

Fig. 7.1: *Trench, ready for the bokashi pre-compost.*

Once you have dug your trench, use the spigot on your bucket (if it has one) to drain off all of the excess bokashi tea that might have accumulated in the bottom and set it aside to be used later. Then pour the pre-composted contents of your bokashi bucket into the trench and spread them out evenly across the width and length of the trench.

Fig. 7.2: *Bokashi pre-compost added to the trench.*

Add an inch or so of soil on top of the pre-compost and then mix the soil thoroughly with the pre-compost using your shovel. Then backfill the trench, leaving about six inches of cover over the mixture, a layer thick enough to keep any curious critters out.

Fig. 7.3: *Bokashi pre-compost mixed with soil.*

Fig. 7.4: *Bokashi pre-compost buried in the trench.*

The soil micro- and macro-flora will then go to work assimilating the fermented pre-compost into the soil. You will be able to plant over and into this area in two weeks, and most of the

Fig. 7.5: *Bokashi pre-compost and soil after two weeks in the soil. Only a few coffee filters and paper towels remain.*

pre-compost will probably be unrecognizable in four weeks. This time frame is an estimate and may vary based upon local soil conditions and the outside air temperature; less active soils and colder temperatures will increase the assimilation time. Very dense scraps such as grape stems and bones may take longer to fully break down.

Fig. 7.6: *Worms from the surrounding soil moved in to eat the pre-compost.*

Isn't bokashi acidic, and won't it increase the acidity of my soil?

Yes and yes, but only temporarily. The bokashi fermentation process does produce organic acids, so the adjacent soil will show a rise in acidity right after the bokashi pre-compost is applied. But within a week the soil web will start to neutralize the acidity and the pH will gradually return to neutral. So long-term applications will not increase the soil acidity.

I often suggest that people who are trying to incorporate a large amount of bokashi pre-compost into a garden plot should work in a grid pattern. Start in an empty corner of the plot and begin trenching and burying in a column, putting each successive set of pre-compost in a trench right next to the previous trench. Work your way vertically through the garden until you come to the opposite border. Then move over and start a new column of trenches next to the previous one, and so on and so forth.

Progressing in this manner has several benefits. First, it helps you keep track of where you buried your last set of scraps. Second, it starts inoculating and building the soil in a systematic manner. By introducing microbes and organic matter to an area, you are providing the conditions for soil life to multiply and flourish. Worms and other soil organisms will start moving into the area to feed on the newly introduced scraps. By slowly expanding out of that area, you are providing the soil life with a constant food source and a favorable soil structure, allowing it to expand throughout your garden. Combining this technique with a cover crop or adding a surface mulch such as straw are easy ways to reinvigorate dead soils.

If you don't have an area of land that can be trenched because of your location or because your land is completely planted out, there is a simple solution; you can introduce bokashi pre-compost to the soil in pockets between existing plantings. Simply dig a twelve-inch-deep hole in between the plants, leaving at least

Fig. 7.7: *Adding bokashi pre-compost in between existing plantings.*

Fig. 7.8: *Simply dig a hole, add the pre-compost, and cover.*

twelve inches between the outer edge of the hole and the nearest plants in all directions; be cognizant of the nearby plants and their roots as you dig. Then add the pre-compost to hole, mix it with some soil, and cover it with at least six inches of soil. Over the next few weeks, the soil biota will feed on the newly introduced organic matter and your plants will have access to a slow-release fertilizer and all of the benefits that go with a good soil structure.

Applying the bokashi pre-compost in individual holes allows you to put the fertilizer directly where it is needed. And unlike traditional soil amendments that are applied dry, the bokashi pre-compost is incorporated wet, so you won't be stealing water from the nearby soil. In addition you will be adding to the long-term water-holding capacity of the soil as the scraps are converted to humus.

If you don't have access to any land that you can trench or dig holes in, you can still use bokashi to turn your food scraps into a usable soil amendment. The answer is to turn the bokashi pre-compost into a soil product that can be used for container gardening or stored for later use. Home composters that live in an area where the ground freezes could also apply this strategy in the winter months and incorporate the finished bokashi soil into their garden when the ground thaws, or use it in their seed-starting mix.

Making bokashi soil is very simple and only involves a few steps. You will need a bucket or similar container that has drainage holes in the bottom, some garden soil or potting soil, and your bokashi pre-compost. It is important to drain away any of the bokashi tea that has accumulated in the bottom of your bucket because, at this stage, excess moisture could cause the bokashi soil mix to go putrid, thereby ruining the process. Don't worry too much because it takes a lot of excess liquid to screw things up. By removing the excess liquid, we are erring on the cautious side because the scraps themselves will still be very high in moisture at this stage.

First, fill the empty bucket one-third deep using the soil you have collected. Next, add the bokashi pre-compost to the bucket until it is about two-thirds full and mix thoroughly. Then add the remainder of the soil so the bucket is completely full. Press the soil down firmly; you don't have to compact it, just apply enough pressure to remove any large air gaps from the mixture. Then cover the full bucket with a piece of plastic or a bucket lid. The cover doesn't have it be airtight at this stage; it just needs to prevent excess moisture from entering or exiting the bucket. Even though the container is covered, you will want to make sure to leave it in an area that is protected from pets or critters to save yourself some unnecessary cleanup down the line. About four weeks later, you should have a usable soil product.

Now wait and let the microbes you added via the soil go to work on the pre-compost. Assuming that the bucket is kept

Fig. 7.9: *The starting ingredients: ⅔ bucket garden soil, empty bucket, bokashi pre-compost.*

Fig. 7.10: *First add ⅓ soil, then add another ⅓ bokashi pre-compost.*

Fig. 7.11: *Soil and bokashi pre-compost mixed thoroughly.*

Fig. 7.12: *Mixture is covered with soil, compacted, then covered for four weeks.*

Fig. 7.13: *Mycelium forming on a bucket of finished pre-compost.*

around 60–80 °F, it should take four weeks for the scraps to break down some more, assimilate into the soil, and rise in pH. Like at the fermentation stage, the warmer the better, since more soil microbes are more active at warmer temperatures. During the finishing process, you may see white mycelium appear on the soil, but don't worry — this is normal and a good sign.

After four weeks, the finished compost is ready to use. You may still see some bits of kitchen waste at this stage, but these are still being fermented and will break down into the soil given more time.

At this stage the finished soil is now ready to use. Treat it just as you would any other high-quality soil mix. You could use the soil as part of a potting mix or incorporate it directly into the garden by topdressing. I like to use this mixture in the bottom of my containers when I upsize transplants. This gives the roots some nutrient-rich organic matter to reach down into, getting the plants off to a good start. If you live in an area where the ground freezes, you can store the finished soil mix until you are ready to use it in the spring. It doesn't matter if you put it inside or outside; it should store for a few months either way without any problems. If it freezes and thaws at this stage, it isn't a big deal.

You can replicate the soil mix finishing process for the bokashi pre-compost in any container and on any scale. Just be sure to mix the bokashi pre-compost at a ratio of at least 1:2 with soil, top with a soil cover, ensure adequate drainage, and wait.

You can also use the same method to rehabilitate used potting soil. Container gardeners regularly find themselves with plant-less, soil-filled containers as the seasons change or plants die off. A lot of these are more dirt than soil, friable and requiring amendments or complete replacement before the next planting. Bokashi pre-compost can be used to add organic matter and nutrients back to these used potting soils, oftentimes in the same container. Just remove two-thirds of the soil from the container and set it aside. Add enough bokashi pre-compost to the container so that it is two-thirds full. Mix the contents thoroughly, top the container off with some of the original soil, and press the contents down firmly. Cover the container to keep excess moisture out and let it sit out of direct sunlight for a month. After this curing phase, the soil can be directly planted into. You don't have to remove the soil or remix it at that point, just plant into it. For small containers or a large number of pots, it is often easier to combine all the soil in one large container such as a trash can. You can then do the mixing and curing in bulk and redistribute the finished soil to the individual containers after it has sat for a month.

People often ask, can I add bokashi pre-compost to a traditional compost pile? The answer is yes, but you are losing some of the advantages of the fermentation process. When you ferment scraps, you are preparing them for rapid assimilation into the soil matrix; you are going straight from fermentation to soil, so adding them to a compost pile is an unnecessary step. Yes, the bacteria in the pre-compost will help inoculate the pile, but the heat and oxygen will ultimately stop them from multiplying and thermophilic bacteria will dominate the composting process. The heat will ultimately cook away some of the nutrients, enzymes, and bacteria in the organic waste, so unless you don't have space to trench or don't want to make a soil mix, don't add the pre-compost to a traditional compost pile. Bokashi pre-compost is best used in the soil, so take advantage of the fermentation process the bokashi composting process initiates to rapidly build up your soils without losing nutrients.

If you compost using worms, you can add bokashi pre-compost to your worm bin. But again, the question comes down to, why? The best, most appropriate use for the bokashi pre-compost is in the soil. So put it there. Why double handle it? There are all sorts of ideas and theories as to why bokashi pre-compost should go into a worm bin, but I don't agree with any of them.

Feed the worms the straight, unfermented food scraps they like to eat and bokashi compost anything else you don't put into the worm bin. When you put the bokashi pre-compost directly into the soil, the worms and soil biota will eat the fermented scraps in situ, giving you castings right where you need and want them. Why add unnecessary steps by going to the worm bin first with bokashi pre-compost? I don't know, but people insist on doing it, so here is how I would approach it.

I would not add the raw bokashi pre-compost straight into the worm bin; it should be finished first. Why? Because the fermented pre-compost is acidic (it will most likely have a pH in the high 3s), and worms don't like that much acidic material. You can greatly reduce the chances of shocking the system by finishing the pre-compost and allowing the pH to rise, thereby making the

Fig. 7.14:
Finishing the pre-compost for the worm bin using equal amounts of shredded paper and pre-compost and a couple of handfuls of worm castings.

material more tolerable to the worms. Finishing the pre-compost is easy and will only take a week.

First, collect equal volumes of bokashi pre-compost (drained of any excess bokashi leachate) and a carbon source such as straw, dried grass, or shredded paper. Mix the two together along with a couple of handfuls of worm castings. Then put the mixture into a plastic bag or bucket, seal it, and let it sit out of direct sunlight at room temperature for a week.

The finishing will trigger some hot composting that will allow the pre-compost to mellow out and return to a more neutral pH. After a week, the mixture can then be incorporated into the worm bin. Just add it to one section of the bin, cover it with some bedding, and let the worms work into it when they are ready. The key to this process is finishing the pre-compost, which will reduce your chances of shocking the system.

That said, there are people out there who add raw unfinished pre-compost directly to worm bins without any observable extreme negative effects. I have experimented with this to try to get a feel for the effects and haven't noticed any big issues. It seems that the worms naturally shy away from the newly introduced raw pre-compost until it has sat for some period of time and they deem it to be OK. I can't give you any more specifics than that, but I would say that the key is to only add a small amount of raw pre-compost at a time, and only to one small section of the worm bin. When I say a small amount, think a handful. That way the acidic effects are minimized and any catastrophic damage can be contained to a particular area.

So if I were to add raw pre-compost to the worm bin, I would choose a corner of the bin, dig a small hole, add a handful of raw pre-compost to the hole, mix it with some of the other bin contents, and cover it. Then I'd watch and wait. This is a key step, as it is important to observe and react to how the worms handle the pre-compost. If everything appears to be going well in a couple of weeks, you can add some more bokashi pre-compost to another section of the bin.

What does going well mean? It means no putrid smells are coming from the bin and worms have entered the area and are consuming the pre-compost. Observe and react. Over time you will start to get a feel for how your worms are handling the compost you are feeding them and can increase the volume of pre-compost you put into the bin based on that. There is no correct method for every situation out there and nothing is cast in stone; it is a process. That is why you don't want to open your worm bin the first time and dump two gallons of bokashi pre-compost onto a four-square-foot area — the risk that it will be too much and you will completely upset the balance of the system you are trying to maintain is just too great. At that point, you won't be able to go back, and the results could be catastrophic.

All of the methods discussed in this chapter have assumed that the bokashi composting process was successful; fermentation occurred as expected and the pre-compost is a good usable final product. But as with any process, things can go wrong along the way.

If you open your bucket after the two-week ferment and see black or blue/green mold on the food waste or smell a rancid odor, something has gone wrong. The contents should have a mildly pleasant sour cider vinegar smell, and the only mold-like substance you see should be white. A few tiny spots of green or black mold on the surface of the scraps aren't a big issue, but a lot of non-white mold is a problem. Again, I want to stress the quantity — an inch of non-white mold here and an inch there isn't serious, especially if the bucket smells OK, and the pre-compost can be used as normal. But a full surface of non-white mold and/ or a rancid odor is a problem.

There are a few factors that could cause this. The most likely causes are that air entered the bucket at some point during the

Fig. 7.15: *A few small patches of blue-green mold. The contents smelled OK, so I treated this pre-compost normally.*

fermentation cycle or too little bran was applied to the scraps when they were added to the bokashi bucket. At this point, you can still save the pre-compost from the trash bin; you just need to apply a different strategy. Dig a twelve-inch trench in the garden, sprinkle a layer of bokashi bran into it, pour the pre-compost on top, add an inch of soil, and mix thoroughly. Then add another layer of bokashi bran and cover everything with six inches of soil and walk away; let the soil web go to work. After one month, the pre-compost should have assimilated into the soil, any noticeably bad odor should be gone, and the soil can be planted into.

If you are space constrained, you could also add the putre-fied contents to an actively managed aerobic compost pile; the thermophilic bacteria and the temperature in the compost pile will take care of them. If neither option is available, then add the putrefied material to your green waste bin with your other yard waste or, as an absolute last resort, to the trash can. Then wash out your bucket to ensure that none of the harmful microbes are introduced to the next batch of food scraps. Take some time to evaluate the process and what might have gone wrong, then adjust the next batch accordingly. As you ferment more and more waste, you will start to get a feel for what works for your situation and what doesn't. More troubleshooting will be covered in the FAQs section at the end of the book.

As you can see, there are numerous ways to use the bokashi pre-compost and to process it into a beneficial soil amendment. Everyone benefits when landowners can improve garden soils and landless apartment dwellers can create productive container soils using their own kitchen waste. This puts less stress on the system and gives people a highly productive soil at home. The finishing methods for bokashi composting are simple, requiring only a few easy-to-find supplies and some time. All the methods described in this chapter required that you drain the bokashi leachate and set it aside. Now let's take a closer look at that leachate.

CHAPTER 8

Bokashi Leachate

IF YOU READ ABOUT BOKASHI ON THE VARIOUS WEBSITES and blogs out there, you will hear a lot about bokashi leachate or tea. This is the liquid generated during fermentation in the composting process. As you start adding organic material to the bokashi bucket, the excess liquid that can't be absorbed by the contents drains down, collecting at the bottom. On its way down, the liquid picks up all sorts of stuff as it passes by and through the organic matter. The resulting tea is a mixture of everything that went into the bucket and then some: water, small bits of organic matter from the kitchen waste, microorganisms from the bokashi bran, microorganisms that came into the bucket on the kitchen waste, the by-products and waste of all of those microorganisms, and more. In most cases, the liquid is orange/yellow in color with a sweet/sour vinegary smell. Its viscosity is somewhere between maple syrup and water, so you should have no problem getting it to flow from a spigot in your bokashi bucket.

When you tap the spigot on the bokashi bucket and collect the leachate, I would suggest using it right away, within hours.

Fig. 8.1: *Bokashi leachate.*

The longer it sits out exposed to the air, the greater the chance it has of spoiling, going rancid, and smelling. If you can't use the leachate right away, I would suggest pouring it down the drain. There is absolutely no harm in that; it isn't a pollutant or toxic, and the microbes in it are beneficial to the waste system, so introducing them to the system is ultimately beneficial downstream. I would not waste my time trying to preserve the leachate by refrigerating it or adding sugar to it. I just don't see a huge benefit there given the relatively small and frequent quantities of leachate that the average bokashi composter will generate. So use it quickly or dump it; you will have more soon enough.

Users with a double-nested bucket system — where the leachate drains from an upper bucket nested into a lower bucket — have to be careful when checking the leachate. In these systems, the user accesses the leachate by lifting the upper bucket up and out of the lower bucket, where the leachate collects. This exposes the leachate to the air, so it is important to collect all of it from the lower bucket. Every time. Even if it is a small amount. If you don't, you run the risk of that leachate going rancid and smelling — smelling really, really bad.

Now that you have the collected the tea, what do you do with it? You can apply the tea to your plants as a biologically active soil amendment. How well does it work? The jury is still out. Some people love it and have had great success using bokashi tea on their plants. I haven't seen any really noticeable effects on my plants, so I don't use it on them. The extra hassle isn't worth the reward for me. I basically collect the tea just to keep the bo-kashi bucket from accumulating too much liquid, which can lead to problems. Then I usually just dilute it and dump it indis-criminately on my lawn. That may seem like a waste, but again, I haven't seen any tremendous effects that justify the extra effort of putting the bokashi tea on my plants. Maybe it is the composition of the bokashi bran I am using, maybe it is the organic waste I am fermenting, or maybe it is something else I can't identify.

That said, I have had numerous people swear by the tea and tell me they have seen great results from applying it to their plants. So it apparently works from some people. And those people are using the same exact bran that I am using, so go figure. If you do try to use the bokashi tea, then worst-case you waste some time; best-case you find something that really helps you out. The sys-tem will produce tea anyway, and you have to dispose of it one way or the other.

So should I dilute the tea or not? The short answer is yes. The long answer is it really depends on your application.

The bokashi tea is going to be acidic, and generally acidity isn't very good for plants, especially in strong concentrations and di-rectly on the foliage. Vinegar makes a great organic "weed" killer after all. So you want to be careful when you apply the bokashi tea to your plants. Test it first on plants you aren't that fond of, and use small doses and get more aggressive if you are OK with the results. Higher order plants such as shrubs and trees should be able to handle a lot more adversity than small herbaceous seedlings. Rotate from plant to plant each time that you use the tea. Even if the results are good, if you keep pouring the tea on the same plant, it may end up accumulating substances it doesn't like or won't like over time.

If you are going to use the tea on your plants, you may want to check its pH periodically to get a handle on what you are actually working with. If the pH is below 6.0, then I would dilute it no matter what. How much? 100 to 1 is safe (a lot of water to a little tea), but it may be overkill. Again, test in small batches and work up from there. In general common sense prevails. Dilute some, rotate through your plants, observe, and adjust. Also, I would shy away from using high concentrations of bokashi leachate directly on foliage. Most plants aren't going to be able to handle the acidity, and you will likely end up burning the leaves. So increase the dilution rate to something like 500 to 1 if you want to put it on the leaves, or just use the bokashi leachate as a soil drench. I would also steer clear of fruiting plants that are in the flowering stage; you don't want to risk applying too high of a concentration of tea and having the flowers drop.

I have done a few unscientific experiments where I have applied bokashi tea in various concentrations to "weeds" on my property. Even at full concentrations, I haven't seen any terribly negative effects. This could be specific to the bokashi tea I am producing, as the tea's composition and pH vary from batch to batch because the feed going into the system isn't the same every time. The tea coming out of my system is acidic, but only weakly; the pH is usually between 4.5 and 6.0. This isn't anywhere near a pH of 2 or 3, which will burn plants for sure, but could cause trouble if you get a batch at 4.5 and apply it straight to the wrong plant. I would still exercise caution and go the dilution route, just it case. I would rather err on the side of diluting more and then adjust the ratio as time goes on than apply full-strength tea and damage my plants, even if the odds of that are low.

If you drain the leachate and it smells bad — think *really* bad — then something went wrong in the fermentation process. The bokashi leachate itself should smell vinegary, not rancid. There are several likely causes of putrid leachate:

1. Was the fermentation vessel truly airtight? Air entering the vessel can allow some aerobic organisms to propagate, causing the contents to putrefy and smell. Check your seals and make sure that everything looks in order, meaning the seals actual seal.
2. Was there too much liquid in the bucket? A large amount of liquid collecting in the bottom of the bucket can go rancid. The easiest ways to avoid this are to drain the bucket every other day and not add a lot of high-liquid contents to the bucket.
3. Moldy or already spoiled food waste was added to the bucket. Unless you compensate this by adding a lot of extra bran and/ or AEM, the bad bacteria will propagate, causing the contents to putrefy. The easiest way to fix this problem is to not add spoiled food waste to the bucket or, if you do, add a lot of extra bran to compensate and change the bacterial balance. The extra bran will introduce more beneficial microbes into the mixture, which will increase their odds of outcompeting the putrefying microbes.
4. There wasn't enough bran added to the bucket regardless of contents. The microbes are transferred to the food waste via the bran, so a lack of bran can result in a lack of microbes to do the fermentation. If the contents aren't fermenting, then they are somewhere between fermentation and putrification, and that will tend to smell bad.

One other important point I want to make is that the production of bokashi tea is not an indicator of the success of the
fermentation process; it is an independent variable. In other
words, lack of tea doesn't mean that fermentation wasn't successful, just as the presence of tea doesn't mean it was successful. If
you put a lot of wet stuff into the bucket, you will tend to get tea,
and if you add stuff that isn't so wet, you will tend to not get tea.
People often ask me things like, "I don't have any tea or I have
very little tea, what I am I doing wrong, why isn't the process
working?" My response is that the production of tea isn't tied to
a successful fermentation, you just aren't adding a lot of liquid or
high-moisture contents to the bucket. And that is perfectly fine.
If your bucket smells OK and is fermenting OK, but you aren't
getting bokashi tea, it's all good.

If you want to increase the amount of tea your buckets produce, you can do things like blend your food waste or pour small
amounts of water through the buckets when they start filling up,
but in my opinion this doesn't seem worth it. Again, the extra
work isn't worth the reward, so why bother?

Bokashi tea is just another by-product of the bokashi composting process. The true value of bokashi composting will
always be the fermented kitchen waste, not the tea. Maybe that
will change in time, but for now the jury is still out on the bokashi
tea.

Conclusion

B OKASHI ISN'T A PANACEA OR A SILVER BULLET. It is just another tool to have in our toolbox so we can combat climate change and the destruction of our environment. Hopefully you can take some or all of the techniques and strategies discussed in this book to help reduce your environmental footprint. The techniques are time tested, fairly straightforward, inexpensive, and effective. The bokashi composting process is an effective outlet for recycling organic waste back into the soil and should be more widely recognized as such, and seen as a viable composting option. It has a lot of benefits over traditional aerobic composting, which gives it a niche.

Bokashi composting is contained and can be done on a small scale, which appeals to a lot of people in small house or apartment situations. It also takes away a lot of the negatives that currently keep many people from composting anything, and gives them another way to process their own kitchen waste into a beneficial soil amendment. If even a small percentage of us kept our kitchen waste on our own property or a property nearby, the impact would be huge. Not only would it keep organic waste out of the landfill but it would dramatically improve local soil conditions — and that is the "gateway drug." If you compost, you are going to want to put your compost to work growing stuff (I don't know very many people who compost but don't grow anything). So if you start composting, you will start growing things. People can use their improved soils to grow edibles, ornamentals, medicinals, or other beneficial plants. This would have a whole variety of beneficial trickle-down effects. Most importantly we would be turning a waste stream into a resource.

I hope that I have also shown that bokashi isn't a hocus-pocus form of composting. There is a legitimate biological process taking place here. Even if you dismiss the theory that adding the microbes in EM to the soil has positive effects, one thing is clear — they break down organic matter quickly and effectively. We are simply using very powerful and effective microbes to do what they evolved to do, in a way that is mutually beneficial. Support for bokashi composting and EM is growing, and there is some grassroots research being conducted out there, mostly by EM manufacturers, bokashi manufacturers, and homebrewers. Numerous cases studies have been conducted around the world using EM and EM bokashi; a small sampling is shown in Appendix B. That said, the research base is far too small, and much more research needs to be done, scientific or otherwise.

The ingredients in the bokashi process aren't genetically modified, synthetic, or laboratory-based chemicals, so that will most likely keep big corporate and government money out of the research space. So like other processes in the alternative science space, the research will come from us, the general public. It will be our efforts and use that ultimately advance the process and application of EM, bokashi, and other natural farming techniques. These techniques are very powerful, not only in their applications but also in their origins. They are simple and homemade, which means that they can be used almost anywhere in this world. This gives power to people and takes it away from the large corporations that control almost all of the chemical alternatives.

Bokashi is just a small piece of the pie. Yet it is an important piece because of its potential impact on the system. And there is so much more in the natural farming space beyond bokashi, and so much of it can be made right in our own garages and kitchens, using basic ingredients and simple techniques. So get curious and start exploring. You have the tools at hand to change your soil and your world, and there is no better place to start than with bokashi.

Appendix A: Bokashi FAQs

I tried fermenting my first bucket of scraps and it didn't work, what happened?

FIRST, KEEP IN MIND that we are dealing with live microorganisms and using them in a real-world, not a laboratory, environment (our bokashi bucket). We are feeding those microorganisms different amounts and types of foods all of the time. And we are introducing those microorganisms to our environment in unknown, varying amounts via the bokashi bran. What I have tried to communicate to people is a process for composting with bokashi, not a strict set of guidelines. I have tried to show what should work and what definitely won't work, leaving room for modification in between and outside. It is important for each user to keep track of their own experiences, take notes, and then react and make changes accordingly.

If you tried to bokashi compost and it didn't work, resulting in a bad batch, don't be discouraged. Something obviously caused the process to go bad. It is important to pinpoint the cause so you can correct it for future batches. Don't just say it didn't work, ask yourself, what went wrong? Was the bucket too wet, were there big chunks of meat, did spoiled food go in, etc. It is important to troubleshoot bad batches and make changes in future batches. In general more bran is better than less, a drier bucket is better than a wetter bucket, and a warmer bucket is better than a colder bucket. More sugary scraps will tend to ferment quicker and easier than less sugary scraps, and small pieces will ferment better than larger pieces.

How do I know if there is too much moisture in the bucket?

If you open the bucket and see drops of condensation on the lid, then the bucket is too moist. Too much moisture can lead to problems. If the contents are too wet, you have a greater chance that they will putrefy and stink.

Add some shredded paper, bread, or other dry material to help soak up some of the moisture. Pay attention to the moisture levels of the materials when you add them to the bucket. For example, if you add a jar of applesauce you will need some dry material to offset the extra moisture. In contrast, if you are adding apple peels and cores to the bucket, you will be fine moisture-wise.

Also be sure to drain your bokashi bucket regularly, ideally every couple of days. Letting the bokashi tea sit for too long at the bottom of the bucket can lead to it spoiling and producing off odors.

How much bokashi bran do I use?

In general you will need to add 2–3 tablespoons (a handful) of bran for every 2 quarts of waste. This equates to a rough dusting over the total surface of a layer of organic waste about two inches thick. The microorganisms in bokashi bran are powerful so it doesn't take a lot, but you do want to make sure that the surface is adequately covered. Keep an eye and a nose on your bucket. If you see green mold, add a little more bran. If you smell strong, rancid, putrid odors, add more bran. If you are composting harder to break down items such as bones, cheese, corn cobs, meat, etc., it is better to apply extra bran to these items when you add them to the bucket.

Where can I keep my bokashi bucket?

For optimum results keep the bucket somewhere warm. Fermentation generally works better in warmer temperatures, over 70°F, so the warmer the location the better. Cooler temperatures will slow down the microbes and thus the fermentation process. Yes, it will still work, but it will take a lot longer due to depressed microbial activity. Buckets should also be kept out of direct sunlight and out of temperatures above 110°F or below 45°F.

The cupboard under the sink is usually the most convenient place to keep a bokashi bucket. The system won't create any nasty smells or attract any insects, so you don't have to worry about keeping it indoors. Like anything else, the best place for the bokashi bucket is the spot where you will actually use it.

How often should I add scraps to the bucket?

In order to successfully ferment the food, the microbes need to be kept as anaerobic as possible, so you don't want to be opening the system every time you generate a little bit of kitchen waste. I suggest adding scraps to the bokashi bucket once every two days. That will give you a chance to accumulate some scraps, while minimizing the amount of air that enters the bucket. If you have a problem with flies or really hot weather causing quick spoilage, you could add your scraps to the bucket every day, but I don't suggest opening it more than once a day. You also want to make sure that you are getting the scraps into the bucket in a timely manner, hence the two-day maximum, to decrease the chances of the scraps spoiling and going moldy outside of the bucket. Moldy scraps can introduce higher concentrations of unfavorable microbes into the system, and this can ultimately lead to the contents going bad (bad molds).

What is the liquid being created and what can I do with it?

The liquid at the bottom of the bucket is a mixture of essential microorganisms, by-products from the fermentation process, and liquids from the wastes you put into the bucket. It contains valuable nutrients, so it shouldn't go to waste. More than likely, this liquid will be acidic, so it is best to dilute it with water at a ratio of at least 100:1 (about a quarter of a cup per gallon) and apply it directly to the soil as a drench. If you don't want to apply it to plants, then feel free to dump it straight down the drain. No need to dilute. The microorganisms will help the septic system and help keep the drainpipes clear.

How often should I drain the liquid?

Excess liquid inside of the bucket will make it go bad and spoil, so it is important to drain the bucket every couple of days, if not every day. Use the liquid immediately, as it will tend to go bad if left exposed to air for long (more than a day, if not less). When it goes bad, it will smell bad. You might be able to avoid spoilage by putting the leachate into a closed bottle and putting that in a fridge, but this may not work, so it is always best to dilute it and use it right away.

How long should I ferment my waste for?

Fermentation generally takes between fourteen and thirty days to complete, depending on the outside temperature, the contents of the bokashi bucket, and the quality of the bokashi bran. On average you can successfully ferment food waste in fourteen days, but it may take longer. Some users leave the bucket for longer periods of time to get a more complete fermentation. Heavier scraps such as bones, avocado skins, and meat will take longer to break down. Fully broken-down organic waste should have a sweetish, sour, or pickled odor and will generally be mushy when squeezed.

Why and when should I compress the waste inside the bucket?

Always. Any time you add new waste to the bucket, compress it. The microorganisms break down the waste using an anaerobic process, so compressing it helps to remove oxygen from the waste mass and increase surface contact between waste particles and the inoculated bokashi bran. This will increase efficiency inside the bucket.

You can compress the waste with a potato masher, your hand, or some such tool. A plastic grocery bag laid on top of the contents works well. It gives you a clean surface to compress the waste and creates an air barrier to keep oxygen out. Some people go as far as putting a brick or a bag of stones on top of the bag to further compress the waste.

Another option for compressing the waste is a homemade pressure plate. A local bokashi user sent me this idea, and it works really well. It is cheap, easy to make, and will save you some hassle over time.

I built this pressure plate using an old five-gallon bucket lid and some scrap wood. An old bucket bottom, a piece of wood, or a piece of scrap plastic could also be used; anything rigid and flat will work. Just make sure to cut the piece of scrap into a circle that has a slightly smaller diameter than the inside diameter of your fermentation bucket. This will allow the pressure plate to fit inside the bucket while leaving as little room as possible between the bucket wall and the edge of the plate. Some buckets are slightly tapered and have a larger diameter at the top than the bottom.

Fig. A.1: *Homemade pressure plate.*

Fig. A.2: *Pressure plate inside a five-gallon bucket.*

Ideally you want the pressure plate to reach down into the lower third of the bucket, so size your lid accordingly.

Then just screw some scrap wood onto the pressure plate to use for support and a handle.

Fig. A.3: *Finished pressure plate. An old bucket lid with wood supports.*

You could make similar pressure plates for non-round commercial bokashi buckets using the same basic idea.

Overall, pressure plates are simple to construct and well worth the effort, saving you hassle and mess down the line as you ferment all your kitchen waste using bokashi.

What do I do with my fermented waste?

Fermented waste can be added directly to the soil, a worm bin, or a compost pile. You can do this by digging a hole or trench, dumping the waste into the hole, and covering it with soil. I suggest waiting at least two weeks before planting anything in that area to make sure that the waste is fully broken down and assimilated into the soil. Bokashi pre-compost should be finished to raise its acidity before adding it to a worm bin.

Animals are digging up my waste. What can I do?

If animals are digging up the waste, then they can smell it, so you need to bury it deeper and cover it with more soil. Animals typically ignore fully fermented food wastes, but if they are really becoming a problem, try burying the waste a foot deep. That should be deep enough to prevent the smell from escaping. You can also try placing a rock or a small sheet of wire mesh over the hole where the waste is buried.

What if I forget to add bokashi bran to the waste?

The bokashi bucket is just an airtight container. You need the bokashi bran, which is inoculated with beneficial microorganisms, in order to start the fermentation process and get the pathogen suppression and the odor-reducing qualities that come with bokashi composting. If you don't add the bokashi bran, you will have bacterial action going on in the bucket, but it will be random. Odds are that it will be the wrong kind — it won't be optimum, the contents will putrefy, and you will have a failed bucket on your hands. And the failed bucket will smell bad.

Bokashi works on the principle of domination, in which a group of symbiotic microorganisms overpower and outnumber harmful microorganisms and keep them at bay. The bokashi bran allows the user to hyper-inoculate food scraps with these beneficial microorganisms, so that they can control the microbes that lead to purification. Without the bokashi bran, you aren't fermenting, you are putrefying. You need the bran.

Can I bury the bokashi pre-compost near plants?

Bokashi pre-compost is acidic, and that means there is a risk it will burn plant roots. To be safe, I suggest burying bokashi pre-compost outside the immediate vicinity of plants, or at least ten inches away. Some plants may be affected by the acid nature of bokashi compost, so err on the side of caution. You also have to consider the plants you would be burying the waste around. If you bury a bucket of waste by a forty-foot black locust tree once a month, it won't be a problem. Bury a bucket every week next to a foot-tall newly planted tomato plant, and it may be a problem; then again it might not.

It isn't an exact science, and every batch of bokashi compost is different. There aren't any right answers that are gospel for each individual situation. Too much of anything too often in one spot may lead to problems. If you want to be extra safe, bury the bokashi pre-compost in a trench in an unused part of the garden, add it to a traditional compost pile, or pre-compost it and add it to a worm bin. All of these actions will allow the pre-compost to return to a more neutral pH and homogenize with the soil better before it comes into direct contact with plant roots.

Can I keep my bucket outside during the winter? I live where it gets cold.

No, it won't work. I am not sure if the microorganisms would die, but they would surely shut down and go dormant. The higher the temperature, the better for fermentation (70–100 °F being optimum); lower temperatures slow down the fermentation process (if it occurs at all). The container itself would hold up as long as it wasn't packed so tight that the freezing contents could expand and crack it. If you want to compost using bokashi and you live in a cold area, you will have to keep your bucket inside your house or another heated area during the winter.

Do I have to turn the bokashi compost like traditional compost?

No, you don't. That is one of the benefits of composting with bokashi; no turning is required. Bokashi is an anaerobic process, so you don't want to introduce extra air into the system, so do not turn the mixture while fermentation is taking place. All you have to do is fill the container, compress and seal it, walk away, and wait.

How should I store my bokashi bran?

The best way to preserve the shelf life of dry bokashi bran is to keep it dry and store it anaerobically out of direct sunlight and away from extreme heat. Stored this way, it should remain viable for up to one year.

Can I compost/ferment/add liquids to my bokashi bucket?

I would not. Extra moisture in the system can lead to problems that result in the bucket going rancid and failing. By adding straight liquid to the bokashi system, you are throwing off the moisture balance. True beverages such as beer, milk, juice, etc., are better off drunk, poured down the drain, or diluted and poured onto the soil.

Why even add the liquids to the bucket in the first place? You aren't going to get that much benefit out of fermenting them in small amounts with the other kitchen waste. Most of them will drain right through the contents and end up in the bottom of the bucket unchanged anyway. In a bokashi bucket without a drain, this can cause some serious issues. In a system with a drain, you end up draining off what you just put in with no obvious benefits, so you are better off just not adding liquids in the first place.

What about semi-liquids? Things like soups, stews, smoothies, cottage cheese? All of these are fine to add to the system. I would take some extra care and mix some dry material with these items if you are adding large quantities of them to your bokashi bucket. I would also add some extra bran to the system as you add these items.

I opened my bokashi bucket and saw green, blue/green, or black mold. What happened?

Colored (non-white) molds can be caused by a few things:

1. Air entered the bucket during fermentation.
2. Some scraps were already spoiled or moldy when they were put into the bucket.
3. Too little bran was applied
4. Some combination of the above.

How can this be corrected for future batches?

1. Ensure that the container is still airtight. A quick and easy way is to empty the container, fill it half full with water, seal it and turn it upside down. If it leaks water, then it will leak air. If water doesn't come out, then you probably do not have an air leak.
2. Do not add moldy scraps to the bokashi bucket. You want the microorganisms in the bokashi bran to dominate any foreign harmful molds. By adding moldy scraps, you give the harmful molds a foothold in the bucket, and they could proliferate and cause problems. If you absolutely have to add moldy scraps to the bucket, I would suggest spraying them (and the mold) with AEM, then adding extra bokashi bran. That should provide enough EM to overpower the mold. But the best practice is to not add any moldy material to the bucket at any time.
3. Add more bran. It may take some time to dial in the appropriate amount of bokashi bran that you need to use for your setup, so observe and react over time. Everyone's scraps, air temperatures, and bokashi bran are different, but in general a dusting on every one-inch layer seems to be work fine. If you are having problems, try adding more bran. It is cheap and easy enough to make, so there is no need to micromanage how much bran you are applying. Err on the side of applying more bran than less.

Appendix B: Case Studies

VARIOUS CASE STUDIES HAVE BEEN CONDUCTED WORLD-WIDE over the last twenty-five years on the effects of EM and EM bokashi anaerobically fermented food waste. In this section, I present a variety of studies that highlight some of the beneficial effects of EM and EM-related products.

SCD Probiotics Application In Composting: Effects of All Seasons Bokashi™ in Soil Quality

SCD Probiotics, 2004

"All Seasons Bokashi™ is classified as an organic fertilizer consisting of beneficial organisms which provide favorable conditions for plant growth. The lactic acid fermentation which takes place because of *Lactobacillus* and yeast presence is the result from the activities of the consortium of microorganisms at an early stage. Lactic acid is also a direct result of low pH, suppressing most undesirable microorganisms. All Seasons Bokashi™ contains a high phosphorous content suggesting it to be a good nutrient source for plants.

Because of anaerobic conditions, aerobic nitrification is suppressed and there remains to be high NH_4+ -N and very low NO_3- -N. In a study (Kato et al., 1997) when All Seasons Bokashi™ was applied to Andosol soil, most NO_3- -N underwent nitrification within 20 days. Rapid nitrification was also detected (Yamanda and Xu, 2000) in the field where All Seasons Bokashi™ was applied. However, rapid nitrification does not transpire in soils that lack nitrifying microbes, thus providing the significance of SCD Probiotics.

Due to the nutrient supply of All Seasons Bokashi™ (Kato et al., 1997), there is a higher rate of growth and photosynthetic activity in plants. These are results compared to chemical fertilizer experiments that are less efficient in providing well developed roots in plants. In fact, enhanced root development and root growth are the most obvious effects of All Seasons Bokashi™ treatments (Yamada et al., 1997). It is even possible that phytohormones or other auxin-type growth regulators produced by SCD Probiotics and present in All Seasons Bokashi™ are responsible for the activation of such root activity (Yamada et al., 1997)."

AgrowTea™ Dilution Study Impact on Wheat Berry Grass

Lawrence R. Green, MD, PhD, Bokashicycle, January 2012

"A pilot study was done to assess the impact AgrowTea™ watering has on Wheat berry grass seedlings.

Wheat berry grass seeds were placed in a container with tap water and allowed to soak overnight. Seedlings were allowed to germinate in six inch pots filled with garden soil to which water and or diluted AgrowTea was added. Leaf length was measured and followed for a period of approximately 2 weeks at which time no further growth was observed. Plants were watered with plain water or AgrowTea™.

Full grown Wheat berry grass bunches were then carefully removed from their pots. Root structure was examined after carefully removing residual soil with cold water washing. It was apparent that even small amounts of AgrowTea™ added to the soil resulted in a high-density root structure.

Numerous lateral shoots were observed directly beneath the soil forming a crown and root matrix most apparent in a stereoscopic microscopic examination and it appears related to the dilution ratio of tea to water. Although the crown and matrix root structure was evident in all pots, it was most evident at a 1:50 (tea to water) dilution ratio and least evident when only water was used.

AgrowTea™ is a liquid extract obtained by fermenting organic waste that is rich in microbes, trace minerals, nutrients and fermented metabolized fibrous debris free of pathogens used in the field to restore microbial flora and organic content. It is the end product of acidic anaerobic (Bokashi) fermenting after the Bokashi bio-pulp (AgrowPulp™) separation is done."

Influence of Method of Application of Effective Microorganism on Growth and Yields of Selected Crops
EMRO, Dr. U.R. Ravi Sangakkara

"Effective Microorganism (EM) solutions have been proven to be very useful and effective in a diverse range of environments, both under field and plant house conditions. The beneficial role of these microbial solutions stems from the ability to break down organic matter, thereby providing plant nutrients, and enhancing soil physical and chemical properties. However, different systems adopt different methods of applying EM ranging from soil or foliar application to composting. But the results of different methods of application could vary significantly. Thus, a field study evaluated different methods of applying EM on growth and yields of two important vegetables, namely French beans and tomato, when grown with two types of organic matter.

Foliar application of EM on two occasions after planting produced the lowest yields. In contrast, applying EM to organic matter added to the field plus two foliar applications produced the highest yield. Yields obtained by the addition of organic matter composed at an external location with EM prior to field application along with two foliar applications were marginally lower. The yields produced by plants supplied with EM in other forms were intermediate to these treatments, irrespective of seasons. The study illustrated the benefits of in situ composting of organic matter with EM and foliar applications at important growth stages to maintain high yields of the selected vegetables."

Effects of Organic Fertilizers and a Microbial Inoculant on Leaf Photosynthesis and Fruit Yield and Quality of Tomato Plants

Hui-lian Xu, Ran Wang, and Md. Amin U. Mridha, 2000

"An experiment was conducted to examine the effects of applications of an organic fertilizer (bokashi), and chicken manure as well as inoculation of a microbial inoculant (commercial name, EM) to bokashi and chicken manure on photosynthesis and fruit yield and quality of tomato plants. EM inoculation to both bokashi and chicken manure increased photosynthesis, fruit yield of tomato plants.

Concentrations of sugars and organic acids were higher in fruit of plants fertilized with bokashi than in fruit of other treatments. Vitamin C concentration was higher in fruit from chicken manure and bokashi plots than in those from chemical fertilizer plots. EM inoculation increased vitamin C concentration in fruit from all fertilization treatments.

It is concluded that both fruit quality and yield could be

significantly increased by EM inoculation to the organic fertilizers and application directly to the soil."

A Pilot Study Comparing Gaseous Emissions Associated with Organic Waste Treated With and Without Bokashi Fermentation

Lawrence R. Green, MD, PhD, Bokashicycle, February 2009

"By simple inspection it is very apparent that the Bokashi fermentation process results in no measureable gas build up within
the fermenter. This was true during the entire 7 day period of
monitoring prior to soil mixing and even after the fermented end
product was placed in the soil mix no gas was produced.

By sharp contrast, within 6 hours of mixing the organic waste
with soil, gas build up in the fermenter was evident. It was also
noted that there was a strong putrid odor when the waste was
examined taken from the non-bokashi fermented container. And
it was clearly evident as waste material even after a week in the
fermenter.

It was also apparent by inspection that the organic waste processed by bokashi fermentation when mixed with the soil and
soil microbes after 7 days in the reactor rapidly lost the appearance of waste material. It appeared virtually identical to the soil
when the fermenter was opened and there was no putrid or foul
odor at the end of the experiment.

Organic waste placed in the soil produces gases that are measurable within 6 hours in a pilot control study experiment and
the putrefaction is evident.

Organic waste processed by Bokashi fermentation produces
no measurable gas during the 7 day fermentation process and
when then mixed with soil it is further degraded without evidence of any gasses being liberated.

Based on these findings it appears that Bokashi fermentation
does not produce measureable gas emissions in its conversion of

organic waste to a nutrient rich end product that can be used to support plants and crops."

Phytophthora Resistance of Tomato Plants Grown with EM Bokashi

Hui-lian Xu, Ran Wang, Md. Amin U. Mridha, and U. Umemura, International Nature Farming Research Center

"In this study, it is found that the easiness of tomato phytophthora infection is associated with nitrogen metabolism and nitrogen status and metabolism in the soil. Tomato plants fertilized with chemical fertilizer contain more nitrate nitrogen and nitrogen compounds such as amino acids than plants fertilized with EM Bokashi. The high concentration of nitrogen compounds might be favourable to the infection and development of phytophthora pathogens. On the contrary, low nitrogen compounds in EM Bokashi fertilized plants might account for the phytophthera resistance. The question is why the nitrate and nitrogen compounds were low in concentration in EM Bokashi fertilized tomato plants. Results showed that the high activity of the enzyme, nitrate reductase, accounted for the low concentration in leaves of EM Bokashi-fertilized plants. This might be due to balance, even release and sustainability of nutrients in EM Bokashi. The activity of dehydrogenase in the soil was also higher in EM Bokashi fertilized plots than in chemical fertilized plots. This might enable nitrogen in soil to be supplied evenly and in different types."

Effects of Effective Microorganisms on the growth of *Brassica rapa*

Yuka Nakano, Brigham Young University of Hawaii, 2007

"The effects of Effective Microorganisms™, one type of microbial fertilizer commercially sold, on the growth of Wisconsin Fast Plants (*Brassica rapa*) and on the microbial density in soil were examined for this study. Two controls and three treatments were prepared, and plant height, diameter of stem, biomass of seed pod and plant, and the population of general microorganisms in soil were measured in order to be compared statistically by ANOVA and Dunnett test. Although there was no significant difference in plant height, treatment with EM bokashi plus EM solution resulted in the thickest diameter of stem, followed by the chemical fertilizer. Both seed pod/plant biomass and microbial density showed a maximum response to the EM bokashi plus EM application. This suggests that EM products are effective for the improvement of growth of *Brassica rapa* and for enhancing the increase of microorganisms in soil."

Investigation on the Properties of EM Bokashi and Development of Its Application Technology

K. Yamada, S. Kato, M. Fujita, Hui-lian Xu, K. Katase and U. Umemura

International Nature Farming Research Center, Hata-machi, Nagano 390-14, Japan.

"Researches were carried-out to elucidate material properties of EM bokashi, the quality estimation methods, and the mechanistic basis for the effects of soil improvement and crop yield promotion. It is known from the analytic results that EM bokashi contains a large amount of propagated *Lactobacillus* and yeast, intermediate substances like organic and amino acids at high concentrations, and 0.1% of mineral N mainly in NH_4 state, and 1% of available phosphorus with a C/N ratio of 10. The quality of EM bokashi depends on water content, additions of molasses and EM. The value of pH can be suggested as an quality criterion of EM bokashi. Effects of EM bokashi on soil fertility and crop growth might result from two different factors, the organic materials and EM microbes with the produced substances. EM application promoted plant growth grain yield and the photosynthetic activity of sweet corn by increasing root development and root activity. It was suggested that in EM bokashi might exist some kinds of microbes and their produced substances stimulating root activity. Further studies are needed to elucidate the related mechanisms."

Observations on the Use of Effective Microorganisms (Kyusei EMTM) on Selected Vegetable Crops Using Nutrient Enriched Water from a Water Recirculated Intensive Fish Production System

J.F. Prinsloo and H.J. Schoonbee, Aquaculture Research Unit, University of the North

"Integrated aquaculture-agriculture production systems were developed primarily for sustainable food production in rural areas of South Africa. The problem of environmental pollution was addressed by utilizing the nutrients in the agricultural waste products to fertilize the fish ponds, thus stimulating pond productivity and releasing nutrients such as nitrogen and phosphorus into irrigation water for vegetable crop production. Due to problems encountered with soil quality as a result of the build-up of nutrients due to chemical fertilizers and the excessive use of pesticides, it was decided to implement EM technology to restore the organic and micro-biological balances of the cultivated lands. Three vegetable crops, namely cabbage, spinach and lettuce, were cultivated under different irrigation systems, using EM, organic compost and inorganic chemical fertilizing programmes. After the first application of EM Bokashi, significant improvements in vegetable yields were recorded. EM treated plots proved to be superior in the yields of cabbage and lettuce clearly exceeding the agricultural average for South Africa. In the case of spinach, EM treated plots under drip irrigation proved to be the most productive. Application of irrigation water as well as soil quality, appeared to be two factors that must be considered when ascertaining the amount of EM to be applied to further improve yields following organic farming. Recommendations are made to evaluate the health status of EM treated soils."

Appendix C: Further Reading

Carandang, Gil A. *Grow Your Own Beneficial Indigenous Micro-organisms and Bionutrients in Natural Farming.* Makati City, Philippines: Bronze Age Media, 2011. Ebook.

Cho, Han-Kyu, and Ju-Young Cho. *Natural Farming: Agriculture Materials.* Seoul, South Korea: Cho Global Natural Farming, 2010.

Higa, Teruo. *An Earth Saving Revolution: A Means to Resolve Our World's Problems Through Effective Microorganisms* (EM). Tokyo, Japan: Sunmark Publishing Inc. 1993. Print.

Hui-lian Xu, James F Parr, and Hiroshi Umemura. *Nature Farming and Microbial Applications.* Boca Raton, FL: CRC Press, 2000. Print.

Appendix D: Works Cited and Notes

The following books and articles are quoted in the book. For page references, see the Notes that follow.

EMRO Japan. "Microorganisms in EM." Web: emrojapan.com: 9000/about-em/microorganisms-in-em.html, accessed April 1, 2013.

EMRO Japan. "What is EM?" Web: emrojapan.com/application/ household1f40.html, accessed April 1, 2013.

Green, L. "A Pilot Study Comparing Gaseous Emissions Associated with Organic Waste Treated with and without Bokashi Fermentation" [Case Study]. 2009. Print

Green, Terrence, and Radu Popa. "Turnover of Carbohydrate-Rich Vegetal Matter during Microaerobic Composting and after Amendment in Soil" [Case Study]. 2011. Print.

Higa, Teruo, and Parr, Dr. James F. "Beneficial and Effective Microorganisms for a Sustainable Agriculture and Environment." p. 1994. Web: agriton.nl/higa.html#Fig1, accessed April 1, 2013.

Inocusol. "*Rhodopseudomonas palustris*, Scientific Classification." Web: enviresol.com/Inocusol/RhodopseudomonasPalustris. htm, accessed April 1, 2013.

Kim, Hong-Lim, Bong-Nam Jung, and Bo-Kyoon Sohn. "Production of Weak Acid by Anaerobic Fermentation of Soil and Antifungal Effect" [Case Study]. 2006. Print.

Le-Hoang, Anh. "Study of *Rhodopseudomonas palustris* toward Photoautotrophic Growth via the Media" [Case Study]. 2008. Print.

Nishio, Michinori. "Microbial Fertilizers in Japan" [Case Study]. 1996. Print.

Pinto, Vinny. "Introduction to Effective Microorganisms (EM)." Web: web.archive.org/web/20051215005439/eminfo.info/moreem1.html#dilutions, accessed April 1, 2013.

Pinto, Vinny. *Fermentation with Syntropic Antioxidative Microbes: An Advanced Guide to Brewing EM Fermented Secondary Products.* 2004. Print.

SCD Probiotics. "SCD Probiotics Inside." Web: scdprobiotics.com/SCD_Probiotics_Inside_s/9.htm, accessed April 1, 2013.

US Environmental Protection Agency. "Types of Composting." Web: epa.gov/compost/types.htm, accessed April 1, 2013.

Notes

1 See epa.gov/compost/types.htm
2 Higa, *An Earth Saving Revolution*, p. 115.
3 Nishio, "Microbial Fertilizers in Japan,".
4 Ibid.
5 Higa, *An Earth Saving Revolution*, p. 51.
6 Ibid, p. 55.
7 Ibid, p. 102.
8 From a lecture.
9 Higa, *An Earth Saving Revolution*, p. 14.

10

10 Pinto, *Fermentation with Syntropic Antioxidate Microbes*, p. 37.

11 Pinto, "Introduction to Effective Microorganisms (EM),".

12 EMRO Japan, "What is EM?"

13 Inocusol, "*Rhodopseudomonas palustris*, Scientific Classification."

14 Pinto, *Fermentation with Syntropic Antioxidate Microbes*, p. 43.

Index

A

acids, in fermentation, 37, 106
activated EM (AEM), 8, 54–57, 63
aerated composting (AC), 11–12, 113
aerobic composting.
    See compost pile.
agricultural microorganisms, 25–26
AgrowTea, 142–143
All Seasons Bokashi, 141–142
ammonia, in compost, 12
anaerobic composting, 12, 29
anaerobic fermentation, 7–8, 13, 29–30, 37
antioxidants, 37, 38

B

beneficial microorganisms (BMO), 4, 8, 14–16
benefits, of bokashi composting, 7–8
blackstrap molasses, 61–62
bokashi bran, 45–72
    storage of, 45, 53, 138
    summary of, 4, 36
    use of, 37, 136
bokashi buckets.
    See also commercial bokashi buckets.
    defined, 4
    with drain, 91–95
    without drain, 96–98
bokashi composting, summary, 1–2, 4

bokashi fermentation, 29–30. *See also* bokashi composting; temperatures.

bokashi leachate, 17, 119–124. *See also* bokashi tea.

bokashi powder. *See* bokashi bran.

bokashi pre-compost summary of, 4, 21 uses for, 101–118

bokashi soil, making of, 108–113

bokashi tea, 73, 78, 95, 130, 132. *See also* spigot buckets.

Brassica rapa, case study, 146–147

browns and greens, ratio of, 9–10

C

Carandang, Gil, 72

carbon dioxide, from waste, 14

carbon substrates, 4, 8, 46. *See also* wheat bran.

carbon-to-nitrogen ratio (C:N), 9–10

chemicals, in agriculture, 25, 27

chloramine, 62

Cho, Han Kyu, 24–25

coffee grounds, 5, 46, 58–59

commercial bokashi buckets, 73–76

compost pile, defined, 5, 7. *See also* aerated composting.

compression, of waste, 87, 93–94, 133–135

container gardening, and pre-compost, 108, 112–113

crop yield, case study, 147–148

D

dairy, in compost, 8–9, 10

Diver, Steve, 29–30

double-nested buckets, 76–81, 120–121

drying, of bran, 53, 63

E

An Earth Saving Revolution, 27, 32–33

effective microorganisms (EM). *See also* EM1. case studies, 143–145, 148–149 components, 38–43

EM bokashi, 27, 30–43

EM1, 26–28, 32, 33. *See also* activated EM.

EM Research Organization (EMRO), 38–39

Environmental Protection Agency (EPA), 9
essential microorganisms (EM). *See also* effective microorganisms; EM bokashi; EM1.
 defined, 5
 microbes in, 15–16, 33–34

F
fermentation vessel, 73–88. *See also* bokashi bucket.
Fermentation with Syntropic Antioxidative Microbes, 57
fermented food waste. *See* anaerobic fermentation; bokashi pre-compost.
final processing. *See* finishing.
finished soil, 108–113
finishing, 5, 108–113. *See also* trench method.
five-gallon buckets, 85–86, 87. *See also* double-nested buckets.
foliar application, 143–144

G
Gamma Seals, 78–81
gaseous emmisions, case study, 145–146
Green, Lawrence, 13
Green, Terrence, 11

greenhouse gasses (GHG), 11–13, 14
Grit Guards, 84
Grow Your Own Beneficial Indigenous Microorganisms and Bionutrients in Natural Farming, 72

H
Han-Kyu, Cho, 72
Higa, Teruo, 15, 16, 25–26, 27, 32–33, 34
history, of bokashi, 23–28
homemade bokashi bran, 45–54
homemade EM, 36, 40, 63–72

I
indigenous microorganisms (IMO), 24. *See also* beneficial microorganisms.
inoculated carbon source, 8. *See also* carbon substrates.
"Introduction to Effective Microorganisms (EM)," 39–40
irrigation systems, case study, 148–149

J
Jung, Bong-Nam, 30
Ju-Young, Cho, 72

K
Kim, Hong-Lim, 30
kitchen waste, bokashi
composting of, 89–98

L
labeling, 50, 95
lactic acid bacteria (LAB), 13,
15, 38–40, 64
lactobacillus serums, 36, 40,
64–72
lids, for bokashi buckets, 73,
74, 80–81. *See also* Gamma
Seals.
liquids, adding of, 138

M
meat, in compost, 8–9, 10, 37,
93, 131, 133
methane, from waste, 12, 14
microbes, in bokashi bran,
14–15, 30
microbial spray, 8
"Microorganisms in EM,"
38–39
moisture levels, in buckets, 17,
108, 124, 130, 138
molasses. *See* blackstrap
molasses.
molds. *See* non-white molds;
white molds.
mother cultures, 33–36, 38,
45, 63–64.

See also activated EM;
essential microorganisms.

N
natural farming, 24–25, 27
Natural Farming, 72
newspaper bokashi bran,
59–60
Nishio, Michinori, 23–24
nitrogen, from waste, 11, 12,
14
non-white molds, 52, 116,
131, 139
nutrients, and fermentation,
13–14, 29–30

O
odors, and composting, 9, 10,
16, 17, 42–43, 116–117
Okada, Mokichi, 24
organic matter, defined, 5–6.
See also kitchen waste.

P
Parr, James, 30
pathogens, 11, 30, 38
pests, and food waste, 10,
23–24, 109, 136
pH
and bokashi, 13, 30, 37, 39,
114, 115
and bokashi tea, 122
photosynthetic bacteria.

See purple non-sulfur bacteria.

phytophthora resistance, case study, 146

"A Pilot Study Comparing Gasesous Emissions," 13

Pinto, Vinny, 36–37, 39–40, 43, 57, 61

plantings, and pre-compost additions, 106–108, 136–137

Popa, Radu, 11

pressure plate, 133–135

"Production of Weak Acid by Anaerobic Fermentation of Soil and Antifungal Effect," 30

ProBio Balance, 35–36

purple non-sulfur bacteria (PNSB), 15, 41–43

putrefied material, 29, 116–118

R

Rhodopseudomonas, 41–42

rice bran, 58

S

scaleability, of composting, 16–17, 86

SCD Probiotics, 35–36, 141–142

Sekai Kyusei Kyo (SKK), 27

sight test, for bran, 52

single-bucket system, 85

size, of compost system. *See* scaleability.

smell test, for bran, 50, 52

Sohn, Bo-Kyoon, 30

soil amendment, 14, 15–16, 101–118, 135–136. *See also* plantings.

species, of microbes, 33–34

spigot buckets, 78, 81–84, 95

splatter guards, 84

sugars, conversion of, 38

T

temperatures, and bokashi processing, 94–95, 131, 137

TeraGanix, 35–36

timelines, for composting, 18–20, 86–87, 132–133

tomato plants, case studies, 144–145, 146

traditional composting, 5, 8, 16, 18, 113. *See also* aerated composting; anaerobic composting; browns and greens; turning.

trench method, 101–105, 106

turning, of compost, 17–18, 137–138

"Turnover of Carbohydrate-Rich Vegetal Matter," 11

U
unit conversions, 54

V
vermicomposting, and
    bokashi, 113–116

W
waste matter, size of, 90–91
water, in composting, 14, 17,
    62

wheat berry grass, case study,
    142–143
wheat bran, 58–59
white molds, 52
white mycelium, 98, 112
worms, and composting.
    See vermicomposting.

Y
yeasts, 15, 41

About the Author

ADAM FOOTER is a permaculture designer with a focus on soil building, food forestry, nitrogen fixation, cover crops, water conservation and harvesting, and natural farming. He is currently converting his suburban property into a functioning homestead using skills from his engineering background combined with permaculture principles. Adam has been researching and promoting bokashi composting for several years as an ideal solution for maximizing the recycling of food waste in any situation. He runs the website www.bokashicomposting.com.

If you have enjoyed *Bokashi Composting,*
you might also enjoy other

BOOKS TO BUILD A NEW SOCIETY

Our books provide positive solutions for people who want to
make a difference. We specialize in:

**Sustainable Living • Green Building • Peak Oil •
Renewable Energy • Environment & Economy Natural
Building & Appropriate Technology Progressive Leadership
• Resistance and Community Educational & Parenting Resources**

New Society Publishers

ENVIRONMENTAL BENEFITS STATEMENT

New Society Publishers has chosen to produce this book on recycled
paper made with **100% post consumer waste,** processed chlorine
free, and old growth free.

For every 5,000 books printed, New Society saves the following
resources:[1]

16	Trees
1,414	Pounds of Solid Waste
1,556	Gallons of Water
2,030	Kilowatt Hours of Electricity
2,571	Pounds of Greenhouse Gases
11	Pounds of HAPs, VOCs, and AOX Combined
4	Cubic Yards of Landfill Space

[1]Environmental benefits are calculated based on research done by the
Environmental Defense Fund and other members of the Paper Task Force who study
the environmental impacts of the paper industry.

For a full list of NSP's titles, please call 1-800-567-6772 *or check out our website* at:

www.newsociety.com

new society
PUBLISHERS